SpringerBriefs in Optimization

SpringerBriefs in Optimization showcases algorithmic and theoretical techniques, case studies, and applications within the broad-based field of optimization. Manuscripts related to the ever-growing applications of optimization in applied mathematics, engineering, medicine, economics, and other applied sciences are encouraged.

More information about this series at http://www.springer.com/series/8918

Vyacheslav V. Chistyakov

From Approximate Variation to Pointwise Selection Principles

 Springer

Vyacheslav V. Chistyakov
Department of Informatics, Mathematics,
and Computer Sciences
National Research University Higher
School of Economics
Nizhny Novgorod, Russia

ISSN 2190-8354   ISSN 2191-575X   (electronic)
SpringerBriefs in Optimization
ISBN 978-3-030-87398-1   ISBN 978-3-030-87399-8   (eBook)
https://doi.org/10.1007/978-3-030-87399-8

Mathematics Subject Classification: Primary: 46N10, 54E40; Secondary: 26A45, 40A30

This Springer imprint is published by the registered company Springer Nature Switzerland AG
The registered company address is: Gewerbestrasse 11, 6330 Cham, Switzerland

To Sveta, Dasha, and Vasilisa

Preface

This book addresses the minimization of special lower semicontinuous functionals over (closed) balls in metric spaces, called the approximate variation. The new notion of approximate variation contains more information about the bounded variation functional and has the following features: the infimum in the definition of approximate variation is not attained in general and the total Jordan variation of a function is obtained by a limiting procedure as a parameter tends to zero. By means of the approximate variation, we are able to characterize regulated functions in a generalized sense and provide powerful compactness tools in the topology of pointwise convergence, conventionally called pointwise selection principles.

The aim of this book is to present a thorough self-contained study of the approximate variation. We illustrate this new notion by a large number of appropriate examples designed specifically for this contribution. Moreover, we elaborate on the state-of-the-art pointwise selection principles applied to functions with values in metric spaces, normed spaces, reflexive Banach spaces, and Hilbert spaces. Although we study the minimization of only one functional (namely, the Jordan variation), the developed methods are of a quite general nature and give a perfect intuition of what properties the minimization procedure of lower semicontinuous functionals over metric balls may have. The content is accessible to students with some background in real and functional analysis, general topology, and measure theory.

The book contains new results that were not published previously in book form. Among them are properties of the approximate variation: semi-additivity, change of variable formula, subtle behavior with respect to uniformly and pointwise convergent sequences of functions, and the behavior on proper metric spaces. These properties are crucial for pointwise selection principles in which the key role is played by the limit superior of the approximate variation. Interestingly, pointwise selection principles may be regular, treating regulated limit functions, and irregular, treating highly irregular functions (e.g., Dirichlet-type functions), in which a significant role is played by Ramsey's Theorem from formal logic.

In order to present our approach in more detail, let X be a metric space with metric d_X (possibly taking infinite values) and $B_\varepsilon[x] = \{y \in X : d_X(x, y) \le \varepsilon\}$ be a closed ball in X of radius $\varepsilon > 0$ centered at $x \in X$. Given a functional $V : X \to [0, \infty]$ on X, not identical to ∞, consider the following family of minimization problems:

$$V_\varepsilon(x) := \inf_{B_\varepsilon[x]} V \equiv \inf \{V(y) : y \in B_\varepsilon[x]\}, \quad \varepsilon > 0, \quad x \in X. \tag{1}$$

In other words, if dom $V = \{y \in X : V(y) < \infty\} \subset X$ is the effective domain of V, then the previous formula can be rewritten, for $\varepsilon > 0$ and $x \in X$, as

$$V_\varepsilon(x) = \inf \{V(y) : V(y) < \infty \text{ and } d_X(x, y) \leq \varepsilon\} \quad (\inf \varnothing := \infty). \tag{2}$$

The family $\{V_\varepsilon\}_{\varepsilon>0}$ of functionals $V_\varepsilon : X \to [0, \infty]$, $\varepsilon > 0$, is said to be the *approximate family of* V. In this generality, the family has only few properties.

Clearly, for each $x \in X$, the function $\varepsilon \mapsto V_\varepsilon(x) : (0, \infty) \to [0, \infty]$ is *nonincreasing* on $(0, \infty)$, and so, we have inequalities (for one-sided limits)

$$V_{\varepsilon+0}(x) \leq V_\varepsilon(x) \leq V_{\varepsilon-0}(x) \quad \text{in } [0, \infty] \text{ for all } \varepsilon > 0$$

and the property (since $x \in B_\varepsilon[x]$ for all $\varepsilon > 0$)

$$\lim_{\varepsilon \to +0} V_\varepsilon(x) = \sup_{\varepsilon>0} V_\varepsilon(x) \leq V(x) \quad \text{in } [0, \infty]. \tag{3}$$

(Note also that if $y \in X$ and $V(y) = 0$, then $V_\varepsilon(x) = 0$ for all $\varepsilon > 0$ and $x \in B_\varepsilon[y]$.)

In this book, in order to have more properties of the family $\{V_\varepsilon\}_{\varepsilon>0}$, we are going to consider the case when $X = M^T$ is the functional space of all functions $x \equiv f : T \to M$ mapping a subset $T \subset \mathbb{R}$ into a metric space (M, d) and equipped with the extended-valued uniform metric $d_X = d_{\infty,T}$ given, for all $f, g \in M^T$, by $d_{\infty,T}(f, g) = \sup_{t \in T} d(f(t), g(t))$. The functional $V : X = M^T \to [0, \infty]$, to be minimized on metric balls in X, is the (Jordan) *variation* of $f \in M^T$ defined, as usual, by[1]

$$V(f) \equiv V(f, T) = \sup \left\{ \sum_{i=1}^{m} d(f(t_i), f(t_{i-1})) : m \in \mathbb{N} \text{ and } t_0 \leq t_1 \leq t_2 \leq \ldots \leq t_m \text{ in } T \right\}.$$

Its effective domain dom V is the space $\mathrm{BV}(T; M)$ of functions of bounded variation, and it has a number of nice properties such as additivity and (sequential) lower semicontinuity, and a Helly-type pointwise selection principle holds in

---

[1] There are a number of interesting functionals on M^T, which can be studied along the "same" lines, e.g., $V(f) \equiv \mathrm{Lip}(f) = \sup \{d(f(s), f(t))/|s - t| : s, t \in T, s \neq t\}$ (the least Lipschitz constant of $f \in M^T$), or functionals of *generalized variation* in the sense of: Wiener–Young [19, 31, 32, 59], Riesz–Medvedev [14–18, 55, 65], Waterman [78, 79], and Schramm [26, 69].

BV$(T; M)$ (see (**V.1**)–(**V.4**) in Chap. 2). Due to this, the approximate family from Eqs. (1) and (2), called the *approximate variation*, is such that

$$\lim_{\varepsilon \to +0} V_\varepsilon(f) = V(f) \quad \text{for all} \quad f \in M^T,$$

and the family of all $f \in M^T$, for which $V_\varepsilon(f)$ is finite for all $\varepsilon > 0$, is exactly the space of all *regulated* functions on T (at least for closed intervals $T = [a, b]$).

Since property (3) holds for the approximate variation, we are able to obtain powerful compactness theorems in the topology of pointwise convergence on M^T generalizing Helly-type pointwise selection principles. As an example, the following holds: *if a sequence of functions $\{f_j\}_{j=1}^\infty$ from M^T is such that the closure in M of the set $\{f_j(t) : j \in \mathbb{N}\}$ is compact for all $t \in T$ and $\limsup_{j\to\infty} V_\varepsilon(f_j)$ is finite for all $\varepsilon > 0$, then $\{f_j\}_{j=1}^\infty$ contains a subsequence, which converges pointwise on T to a bounded regulated function from M^T* (cf. Theorem 4.1).

Additional details on the aim of this book are presented in Chap. 1.

The plan of the exposition is as follows. In Chap. 1, we review a number of well-known pointwise selection theorems in various contexts and note explicitly that there is a certain relationship between such theorems and characterizations of regulated functions. In Chap. 2, we define the notion of approximate variation, study its properties, and show that its behavior is different on proper and improper underlying metric spaces. Chap. 3 is devoted to elaborating a large number of examples of approximate variations for functions with values in metric and normed spaces. Finally, in Chap. 4 we present our main results concerning regular and irregular pointwise selection principles illustrated by appropriate examples.

Acknowledgments

It is a pleasure to express my gratitude to Panos M. Pardalos (University of Florida, USA) for inviting me to publish with Springer. I would like to thank Marek Balcerzak (Łódź, Poland), Patrick Muldowney (Londonderry, Northern Ireland), and Sergey K. Vodop'yanov (Novosibirsk, Russia) for stimulating discussions on the results of this work. The constant kind attention of Springer's Executive Editor, Elizabeth Loew, is greatly acknowledged.

This book was prepared within the framework of the Academic Fund Program at the National Research University Higher School of Economics (HSE) in 2017–2018 (grant number 17-01-0050) and by the Russian Academic Excellence Project "5–100."

Nizhny Novgorod, Russian Federation
July 2021

Vyacheslav V. Chistyakov

Contents

Acronyms

$\mathbb{R} = (-\infty, \infty)$	Set of all real numbers; ∞ means $+\infty$
$[a, b]$	Closed interval of real numbers
(a, b)	Open interval of real numbers
$(0, \infty)$	Set of all positive real numbers
$[0, \infty) = \mathbb{R}^+$	Set of all nonnegative real numbers
$[0, \infty]$	Set of extended nonnegative numbers $[0, \infty) \cup \{\infty\}$
$\mathbb{N}, \mathbb{Z}, \mathbb{Q}, \mathbb{C}$	Sets of all natural, integer, rational, and complex numbers
$\varnothing$	Empty set; $\inf \varnothing = \infty$, $\sup \varnothing = 0$
$\equiv$	Identical to; equality by the definition

Sets of functions from $T \subset \mathbb{R}$ into a metric space (M, d):

M^T	Set of all functions $f : T \to M$ (p. 7)
$B(T; M)$	Set of bounded functions (p. 7)
$BV(T; M)$	Set of functions of bounded Jordan variation (p. 8)
$\text{Reg}(T; M)$	Set of regulated functions (p. 9 and p. 45)
$\text{St}(I; M)$	Set of step functions (p. 9)
$\text{Mon}(T; \mathbb{R})$	Set of bounded nondecreasing functions $f : T \to \mathbb{R}$ (p. 45)

Chapter 1
Introduction: regulated functions and selection principles

A pointwise selection principle is a statement which asserts that under certain specified assumptions on a given sequence of functions $f_j : T \to M$ ($j \in \mathbb{N}$), their domain T and range M, the sequence admits a subsequence converging in (the topology of) M pointwise (=everywhere) on the set T; in other words, this is a compactness theorem in the topology of pointwise convergence. Our intention here is twofold: first, to draw attention to a conjunction of pointwise selection principles and characterizations of regulated functions (cf. also [34]) and, second, to exhibit the main goal of this contribution.

To be specific, we let $T = I = [a, b]$ be a closed interval in $\mathbb{R}$ and $M = \mathbb{R}$, and denote by: $\mathbb{R}^I$ the set of *all* functions mapping I into $\mathbb{R}$, Mon(I) the set of *monotone* functions, BV(I) the set of functions of *bounded* (Jordan) *variation*, and Reg(I) the set of *regulated* functions from $\mathbb{R}^I$. Recall that $f \in \mathbb{R}^I$ is regulated provided the left limit $f(t - 0) \in \mathbb{R}$ exists at each point $a < t \leq b$ and the right limit $f(t + 0) \in \mathbb{R}$ exists at each point $a \leq t < b$. Clearly, Mon(I) $\subset$ BV(I) $\subset$ Reg(I), and it is well known that each function from Reg(I) is bounded, has a finite or countable set of discontinuity points, and is the uniform limit of a sequence of step functions on I. Scalar- (and vector-) valued regulated functions are of importance in various branches of analysis, e.g., the theory of convergence of Fourier series, stochastic processes, Riemann– and Lebesgue–Stieltjes integrals, generalized ordinary differential equations, impulse controls, analysis in modular spaces [1–5, 27, 39, 40, 43, 44, 48, 51–53, 62, 64–67, 70–73, 76, 77].

In order for a sequence of functions $\{f_j\} \subset \mathbb{R}^I$ to have a pointwise convergent subsequence, it is quite natural, by virtue of Bolzano–Weierstrass' theorem (viz., a bounded sequence in $\mathbb{R}$ admits a convergent subsequence), that $\{f_j\}$ should be *pointwise bounded* (i.e., $\sup_{j \in \mathbb{N}} |f_j(t)| < \infty$ for all $t \in I$). However, a pointwise (or even uniformly) bounded sequence of functions $\{f_j\} \subset \mathbb{R}^I$ need not have a pointwise convergent subsequence: a traditional example is the sequence $f_j(t) = \sin(jt)$ for $j \in \mathbb{N}$ and $t \in I = [0, 2\pi]$ (for more details see Remark 4.8 below). So, additional assumptions on $\{f_j\}$ are to be imposed.

© The Author(s), under exclusive license to Springer Nature Switzerland AG 2021
V. V. Chistyakov, *From Approximate Variation to Pointwise Selection Principles*,
SpringerBriefs in Optimization, https://doi.org/10.1007/978-3-030-87399-8_1

The historically first pointwise selection principles are due to Helly [49]: *a uniformly bounded sequence of functions $\{f_j\} \subset \text{Mon}(I)$ contains a pointwise convergent subsequence* (whose pointwise limit belongs to $\text{Mon}(I)$). This theorem, a selection principle for monotone functions, is based on and extends Bolzano–Weierstrass' theorem and implies one more Helly's selection principle for functions of bounded variation: *a pointwise bounded sequence of functions $\{f_j\} \subset \mathbb{R}^I$ satisfying*

$$\sup_{j \in \mathbb{N}} V(f_j) < \infty, \quad \text{where } V(f_j) \text{ is Jordan's variation of } f_j, \tag{1.1}$$

contains a pointwise convergent subsequence (with the pointwise limit belonging to $\text{BV}(I)$). Note that condition (1.1) of uniform boundedness of variations may be replaced by $\limsup_{j \to \infty} V(f_j) < \infty$ (which is only seemingly more general).

It is well known that Helly's selection principles play a significant role in analysis (e.g., [51, 62, 67]). A vast literature already exists concerning generalizations of Helly's principles for various classes of functions ([5–7, 12–23, 27, 29–34, 42–46, 56, 59, 69, 74–77, 79], and references therein) and their applications [5, 18, 27, 28, 35–37, 42, 48, 50, 70]. We recall some of these generalizations, which are relevant for our purposes.

Let $\varphi : [0, \infty) \to [0, \infty)$ be a nondecreasing continuous function such that $\varphi(0) = 0$, $\varphi(u) > 0$ for $u > 0$, and $\varphi(u) \to \infty$ as $u \to \infty$. We say that $f \in \mathbb{R}^I$ is a function *of bounded φ-variation* on I (in the sense of Wiener and Young) and write $f \in \text{BV}_\varphi(I)$ if the following quantity, called the *φ-variation* of f, is finite:

$$V_\varphi(f) = \sup \left\{ \sum_{i=1}^{n} \varphi\big(|f(I_i)|\big) : n \in \mathbb{N} \text{ and } \{I_i\}_1^n \prec I \right\},$$

where the notation $\{I_i\}_1^n \prec I$ stands for a non-ordered collection of n non-overlapping intervals $I_i = [a_i, b_i] \subset I$ and $|f(I_i)| = |f(b_i) - f(a_i)|$ is the increment of f on I_i, $i = 1, \ldots, n$. (In particular, if $\varphi(u) = u$, we have $V_\varphi(f) = V(f)$.) It was shown by Musielak and Orlicz [59] that $\text{BV}_\varphi(I) \subset \text{Reg}(I)$, and if φ is additionally convex and $\varphi'(0) \equiv \lim_{u \to +0} \varphi(u)/u = 0$, then $\text{BV}(I)$ is a proper subset of $\text{BV}_\varphi(I)$. Goffman et al. [47] characterized the set $\text{Reg}(I)$ as follows: if $f \in \text{Reg}(I)$ and $\min\{f(t - 0), f(t + 0)\} \leq f(t) \leq \max\{f(t - 0), f(t + 0)\}$ at each point $t \in I$ of discontinuity of f, then there is a convex function φ (as above) with $\varphi'(0) = 0$ such that $f \in \text{BV}_\varphi(I)$. A generalization of Helly's theorem for BV functions, the so-called *Helly-type selection principle*, was established in [59], where condition (1.1) was replaced by $\sup_{j \in \mathbb{N}} V_\varphi(f_j) < \infty$.

One more Helly-type selection principle is due to Waterman [79], who replaced condition (1.1) by $\sup_{j \in \mathbb{N}} V_\Lambda(f_j) < \infty$, where $V_\Lambda(f)$ is the Waterman Λ-*variation* of $f \in \mathbb{R}^I$ defined by Waterman [78]

$$V_\Lambda(f) = \sup \left\{ \sum_{i=1}^{n} \frac{|f(I_i)|}{\lambda_i} : n \in \mathbb{N} \text{ and } \{I_i\}_1^n \prec I \right\};$$

here $\Lambda = \{\lambda_i\}_{i=1}^\infty$ is a *Waterman sequence*, i.e., $\Lambda \subset (0, \infty)$ is nondecreasing, unbounded, and $\sum_{i=1}^\infty 1/\lambda_i = \infty$. (Formally, $V_\Lambda(f) = V(f)$ for $\lambda_i = 1, i \in \mathbb{N}$.) For the set $\Lambda \mathrm{BV}(I) = \{f \in \mathbb{R}^I : V_\Lambda(f) < \infty\}$ of functions *of Λ-bounded variation*, Waterman [78] showed that $\Lambda \mathrm{BV}(I) \subset \mathrm{Reg}(I)$, and $\mathrm{BV}(I)$ is a proper subset of $\Lambda \mathrm{BV}(I)$. Perlman [61] proved that $\mathrm{BV}(I) = \bigcap_\Lambda \Lambda \mathrm{BV}(I)$ and obtained the following characterization of regulated functions:

$$\mathrm{Reg}(I) = \bigcup_\Lambda \Lambda \mathrm{BV}(I),$$

where the intersection and the union above are taken over all Waterman sequences Λ (but not over any countable collection of such sequences Λ).

Taking into account that the sets $\mathrm{Mon}(I)$, $\mathrm{BV}(I)$, $\mathrm{BV}_\varphi(I)$, and $\Lambda \mathrm{BV}(I)$ are contained in $\mathrm{Reg}(I)$, Helly's selection principles and their generalizations alluded to above are compactness theorems in the class of regulated functions.

In the literature, there are characterizations of the set $\mathrm{Reg}(I)$, which do not rely on notions of bounded (or generalized bounded) variations of any kind. One of them was given by Chanturiya [9, 10] in the form

$$\mathrm{Reg}(I) = \{f \in \mathbb{R}^I : v_n(f) = o(n)\},$$

where E. Landau's small "o" means, as usual, that $o(n)/n \to 0$ as $n \to \infty$, and the sequence $\{v_n(f)\}_{n=1}^\infty \subset [0, \infty]$, called the *modulus of variation* of f, is defined by ([9], cf. also [48, Section 11.3.7])

$$v_n(f) = \sup\left\{\sum_{i=1}^n |f(I_i)| : \{I_i\}_1^n \prec I\right\}, \quad n \in \mathbb{N}.$$

Note that $v_n(f) \le V(f)$ for all $n \in \mathbb{N}$ and $v_n(f) \to V(f)$ as $n \to \infty$. The author [20, 21] replaced condition (1.1) by (a very weak one)

$$\limsup_{j \to \infty} v_n(f_j) = o(n) \tag{1.2}$$

and obtained a Helly-type pointwise selection principle (in which the pointwise limit of the extracted subsequence of $\{f_j\}$ belongs to $\mathrm{Reg}(I)$ and) which contains, as particular cases, all the above Helly-type selection principles and many others [22, 23, 29]. Assumption (1.2) is applicable to sequences of nonregulated functions, so the corresponding Helly-type pointwise selection principle under (1.2) is already outside the scope of regulated functions. To see this, let $\mathscr{D} \in \mathbb{R}^I$ be the Dirichlet function on $I = [0, 1]$ (i.e., $\mathscr{D}(t) = 1$ if $t \in I$ is rational, and $\mathscr{D}(t) = 0$ otherwise) and $f_j(t) = \mathscr{D}(t)/j$ for $j \in \mathbb{N}$ and $t \in I$. We have $f_j \notin \mathrm{Reg}(I)$ and $v_n(f_j) = n/j$ for all $j, n \in \mathbb{N}$, so (1.2) is satisfied while (1.1) is not (for any kinds of generalized variations including V_φ and V_Λ). A special feature of condition (1.2) is that, for $f \in \mathrm{Reg}(I)$, it is *necessary* for the uniform convergence of $\{f_j\}$ to f, and "almost necessary" for the pointwise convergence of $\{f_j\}$ to f—note that this is not at all the case for (uniform) conditions of the form (1.1).

Dudley and Norvaiša [42, Part III, Section 2] presented the following characterization of regulated functions:

$$\mathrm{Reg}(I) = \{f \in \mathbb{R}^I : N_\varepsilon(f) < \infty \text{ for all } \varepsilon > 0\},$$

where the (untitled) quantity $N_\varepsilon(f) \in \{0\} \cup \mathbb{N} \cup \{\infty\}$ for $f \in \mathbb{R}^I$ is given by

$$N_\varepsilon(f) = \sup\left\{n \in \mathbb{N} : \text{there is } \{I_i\}_1^n \prec I \text{ such that } \min_{1 \le i \le n} |f(I_i)| > \varepsilon\right\}, \quad \varepsilon > 0$$

(with $\sup \varnothing = 0$). They established a Helly-type pointwise selection principle in the class $\mathrm{Reg}(I)$ by replacing (1.1) with $\sup_{j \in \mathbb{N}} N_\varepsilon(f_j) < \infty$ for all $\varepsilon > 0$. In a series of papers by the author, Maniscalco and Tretyachenko ([27, Chapter 5], [34, 74, 75]), it was shown that we get a more powerful selection principle (outside the scope of regulated functions) if (1.1) is replaced by

$$\limsup_{j \to \infty} N_\varepsilon(f_j) < \infty \text{ for all } \varepsilon > 0. \tag{1.3}$$

If we let the sequence of nonregulated functions $f_j(t) = \mathscr{D}(t)/j$ be as above, we find $N_\varepsilon(f_j) = \infty$ if $j < 1/\varepsilon$ and $N_\varepsilon(f_j) = 0$ if $j \ge 1/\varepsilon$, and so, condition (1.3) is satisfied. Moreover, (1.3) is *necessary* for the uniform convergence and "almost necessary" for the pointwise convergence of $\{f_j\}$ to $f \in \mathrm{Reg}(I)$. A comparison of different Helly-type pointwise selection principles is presented in [20–23, 27, 29, 54].

Essential for this work, one more characterization of regulated functions is due to Fraňková [43] (more on regulated functions see also [44]):

$$\mathrm{Reg}(I) = \{f \in \mathbb{R}^I : V_\varepsilon(f) < \infty \text{ for all } \varepsilon > 0\},$$

where the ε-*variation* $V_\varepsilon(f)$ of $f \in \mathbb{R}^I$ is defined by ([43, Definition 3.2])

$$V_\varepsilon(f) = \inf\left\{V(g) : g \in \mathrm{BV}(I) \text{ and } |f(t) - g(t)| \le \varepsilon \text{ for all } t \in I\right\}, \quad \varepsilon > 0$$

(with $\inf \varnothing = \infty$). She established a Helly-type selection principle in the class $\mathrm{Reg}(I)$ under the assumption of uniform boundedness of ε-variations $\sup_{j \in \mathbb{N}} V_\varepsilon(f_j) < \infty$ for all $\varepsilon > 0$ in place of (1.1). However, following the "philosophy" of (1.2) and (1.3), a weaker condition, replacing (1.1), is of the form

$$\limsup_{j \to \infty} V_\varepsilon(f_j) < \infty \text{ for all } \varepsilon > 0. \tag{1.4}$$

Making use of (1.4), the author and Chistyakova [29] proved a Helly-type pointwise selection principle outside the scope of regulated functions by showing that (1.4) implies (1.2). If the sequence $f_j(t) = \mathscr{D}(t)/j$ is as above, we get $V_\varepsilon(f_j) = \infty$ if $j < 1/(2\varepsilon)$ and $V_\varepsilon(f_j) = 0$ if $j \geq 1/(2\varepsilon)$, and so, (1.4) is fulfilled while the uniform ε-variations are unbounded for $0 < \varepsilon < 1/2$.

In this work, we present a direct proof of a Helly-type pointwise selection principle under (1.4), not relying on (1.2), and show that condition (1.4) is necessary for the uniform convergence and "almost necessary" for the pointwise convergence of $\{f_j\}$ to $f \in \text{Reg}(I)$ (cf. also Remark 4.6 below).

All the above pointwise selection principles are based on the Helly selection theorem for monotone functions. A different kind of a pointwise selection principle, basing on Ramsey's theorem from formal logic [63], was given by Schrader [68]. In order to recall it, we introduce a notation: given a sign-changing function $f \in \mathbb{R}^I$, we denote by $\mathscr{P}(f)$ the set of all finite collections of points $\{t_i\}_{i=1}^n \subset I$ with $n \in \mathbb{N}$ such that $t_1 < t_2 < \cdots < t_n$ and either $(-1)^i f(t_i) > 0$ for all $i = 1, \ldots, n$, or $(-1)^i f(t_i) < 0$ for all $i = 1, \ldots, n$, or $(-1)^i f(t_i) = 0$ for all $i = 1, \ldots, n$. The quantity

$$\mathscr{T}(f) = \sup\left\{\sum_{i=1}^n |f(t_i)| : n \in \mathbb{N} \text{ and } \{t_i\}_{i=1}^n \in \mathscr{P}(f)\right\}$$

is said to be Schrader's *oscillation* of f on I; if f is nonnegative on I or f is nonpositive on I, we set $\mathscr{T}(f) = \sup_{t \in I} |f(t)|$. Schrader proved that *if $\{f_j\} \subset \mathbb{R}^I$ is such that* $\sup_{j,k \in \mathbb{N}} \mathscr{T}(f_j - f_k) < \infty$, *then $\{f_j\}$ contains a subsequence, which converges everywhere on I.* This is an *irregular* pointwise selection principle in the sense that, although the sequence $\{f_j\}$ satisfying Schrader's condition is pointwise bounded on I, we cannot infer any "regularity" properties of the (pointwise) limit function (e.g., it may be applied to the sequence $f_j(t) = (-1)^j \mathscr{D}(t)$ for $j \in \mathbb{N}$ and $t \in [0, 1]$). Maniscalco [54] proved that Schrader's assumption and condition (1.2) are independent (in the sense that they produce different pointwise selection principles). Extensions of Schrader's result are presented in [33, 34, 38, 41].

One of the goals of this book is to obtain irregular pointwise selection principles in terms of Fraňková's ε-variations $V_\varepsilon(f)$ (see Sect. 4.5).

This work is a thorough self-contained study of the *approximate variation*, i.e., the family $\{V_\varepsilon(f)\}_{\varepsilon>0}$ for functions $f : T \to M$ mapping a nonempty subset T of $\mathbb{R}$ into a metric space (M, d). We develop a number of pointwise (and almost everywhere) selection principles, including irregular ones, for sequences of functions with values in metric spaces, normed spaces, and reflexive separable Banach spaces. All assertions and their sharpness are illustrated by concrete examples. The plan of the exposition can be clearly seen from the Contents. Finally, it is to be noted that, besides powerful selection principles, based on ε-variations, the notion of approximate variation gives a nice and highly nontrivial example of a *metric modular* in the sense of the author [24–27], or a classical *modular* in the sense of Musielak-Orlicz [58, 60] if $(M, \|\cdot\|)$ is a normed linear space. Results corresponding to the modular aspects of the approximate variation will be published elsewhere.

Chapter 2
The approximate variation and its properties

2.1 Notation and terminology

We begin by introducing notations and the terminology which will be used throughout this book.

Let T be a nonempty set (in the sequel, $T \subset \mathbb{R}$), (M, d) be a metric space with metric d, and M^T be the set of all functions $f : T \to M$ mapping T into M. The set M^T is equipped with the (extended-valued) *uniform metric*

$$d_{\infty, T}(f, g) = \sup_{t \in T} d(f(t), g(t)), \qquad f, g \in M^T.$$

The letter c will stand, as a rule, for a *constant* function $c \in M^T$ (sometimes identified with $c \in M$).

The *oscillation* of a function $f \in M^T$ on the set T is the quantity[1]

$$|f(T)| \equiv |f(T)|_d = \sup_{s, t \in T} d(f(s), f(t)) \in [0, \infty],$$

also known as the *diameter of the image* $f(T) = \{f(t) : t \in T\} \subset M$. We denote by

$$B(T; M) = \{f \in M^T : |f(T)| < \infty\}$$

the set of all *bounded functions* from T into M.

---

[1] The notation for the oscillation $|f(T)|$ should not be confused with the notation for the increment $|f(I_i)| = |f(b_i) - f(a_i)|$ from p. 2, the latter being used only in the Introduction (Chap. 1).

© The Author(s), under exclusive license to Springer Nature Switzerland AG 2021
V. V. Chistyakov, *From Approximate Variation to Pointwise Selection Principles*,
SpringerBriefs in Optimization, https://doi.org/10.1007/978-3-030-87399-8_2

Given $f, g \in M^T$ and $s, t \in T$, by the triangle inequality for d, we find

$$d_{\infty,T}(f, g) \le |f(T)| + d(f(t), g(t)) + |g(T)| \tag{2.1}$$

and

$$d(f(s), f(t)) \le d(g(s), g(t)) + 2d_{\infty,T}(f, g); \tag{2.2}$$

the definition of the oscillation and inequality (2.2) imply

$$|f(T)| \le |g(T)| + 2d_{\infty,T}(f, g). \tag{2.3}$$

Clearly (by (2.1) and (2.3)), $d_{\infty,T}(f, g) < \infty$ for all $f, g \in \mathrm{B}(T; M)$ and, for any *constant* function $c \in M^T$, $\mathrm{B}(T; M) = \{f \in M^T : d_{\infty,T}(f, c) < \infty\}$.

For a sequence of functions $\{f_j\} \equiv \{f_j\}_{j=1}^{\infty} \subset M^T$ and $f \in M^T$, we write:

(a) $f_j \to f$ on T to denote the *pointwise (= everywhere) convergence* of $\{f_j\}$ to f on T (that is, $\lim_{j \to \infty} d(f_j(t), f(t)) = 0$ for all $t \in T$);

(b) $f_j \rightrightarrows f$ on T to denote the *uniform convergence* of $\{f_j\}$ to f on T, that is, $\lim_{j \to \infty} d_{\infty,T}(f_j, f) = 0$. (Clearly, (b) implies (a), but not vice versa.)

Recall that a sequence of functions $\{f_j\} \subset M^T$ is said to be *pointwise relatively compact* on T provided the closure in M of the set $\{f_j(t) : j \in \mathbb{N}\}$ is compact for all $t \in T$.

From now on, we suppose that T is a (nonempty) subset of the reals $\mathbb{R}$.

The (Jordan) *variation* of $f \in M^T$ is the quantity (e.g., [72, Chapter 4, Section 9])

$$V(f, T) = \sup_P \sum_{i=1}^{m} d(f(t_i), f(t_{i-1})) \in [0, \infty],$$

where the supremum is taken over all partitions P of T, i.e., $m \in \mathbb{N}$ and $P = \{t_i\}_{i=0}^{m} \subset T$ such that $t_{i-1} \le t_i$ for all $i = 1, 2, \ldots, m$. We denote by

$$\mathrm{BV}(T; M) = \{f \in M^T : V(f, T) < \infty\}$$

the set of all *functions of bounded variation* from T into M.

The following four basic properties of the functional $V = V(\cdot, \cdot)$ are well known. Given $f \in M^T$, we have:

(V.1) $|f(T)| \le V(f, T)$ (and so, $\mathrm{BV}(T; M) \subset \mathrm{B}(T; M)$);

(V.2) $V(f, T) = V(f, T \cap (-\infty, t]) + V(f, T \cap [t, \infty))$ for all $t \in T$ (*additivity of V in the second variable*, cf. [11, 12, 72]);

(V.3) if $\{f_j\} \subset M^T$ and $f_j \to f$ on T, then $V(f, T) \leq \liminf_{j \to \infty} V(f_j, T)$ (sequential *lower semicontinuity* of V in the first variable, cf. [12, 13]);

(V.4) a pointwise relatively compact sequence of functions $\{f_j\} \subset M^T$ satisfying condition $\sup_{j \in \mathbb{N}} V(f_j, T) < \infty$ contains a subsequence, which converges pointwise on T to a function $f \in BV(T; M)$ (Helly-type *pointwise selection principle*, cf. [6, 18]).

In what follows, the letter I denotes a closed interval $I = [a, b]$ with the endpoints $a, b \in \mathbb{R}$, $a < b$.

Now, we recall the notion of a regulated function (introduced in [2] for real valued functions). We say [20] that a function $f \in M^I$ is *regulated* (or *proper*, or *simple*) and write $f \in \mathrm{Reg}(I; M)$ if it satisfies the Cauchy condition at every point of $I = [a, b]$, i.e., $d(f(s), f(t)) \to 0$ as $I \ni s, t \to \tau - 0$ for each $a < \tau \leq b$, and $d(f(s), f(t)) \to 0$ as $I \ni s, t \to \tau' + 0$ for each $a \leq \tau' < b$. It is well known (e.g., [4, 11, 20, 43, 52, 72]) that

$$BV(I; M) \subset \mathrm{Reg}(I; M) \subset B(I; M),$$

the set $\mathrm{Reg}(I; M)$ of all regulated functions is closed with respect to the uniform convergence, and the pair $(\mathrm{Reg}(I; M), d_{\infty, I})$ is a complete metric space provided (M, d) is complete (see also [29, Theorem 2] for some generalization). Furthermore, if (M, d) is complete, then, by Cauchy's criterion, we have: $f \in \mathrm{Reg}(I; M)$ if and only if the left limit $f(\tau - 0) \in M$ exists at each point $a < \tau \leq b$ (meaning that $d(f(t), f(\tau - 0)) \to 0$ as $I \ni t \to \tau - 0$), and the right limit $f(\tau' + 0) \in M$ exists at each point $a \leq \tau' < b$ (i.e., $d(f(t), f(\tau' + 0)) \to 0$ as $I \ni t \to \tau' + 0$).

Regulated functions can be uniformly approximated by step functions (see (2.5)) as follows. Recall that $f \in M^I$ is said to be a *step function* (in symbols, $f \in \mathrm{St}(I; M)$) provided, for some $m \in \mathbb{N}$, there exists a partition $a = t_0 < t_1 < t_2 < \cdots < t_{m-1} < t_m = b$ of $I = [a, b]$ such that f takes a constant value on each (open) interval (t_{i-1}, t_i), $i = 1, 2, \ldots, m$. Clearly,

$$\mathrm{St}(I; M) \subset BV(I; M). \tag{2.4}$$

Furthermore (cf. [40, (7.6.1)]), we have

$$\mathrm{Reg}(I; M) = \{f \in M^I : \text{there is } \{f_j\} \subset \mathrm{St}(I; M) \text{ such that } f_j \rightrightarrows f \text{ on } I\} \tag{2.5}$$

(if, in addition, $f \in BV(I; M)$, then $\{f_j\} \subset \mathrm{St}(I; M)$ can be chosen such that $f_j \rightrightarrows f$ on I and $V(f_j, I) \leq V(f, I)$ for all $j \in \mathbb{N}$, cf. [11, Section 1.27]).

2.2 Definition of approximate variation

Definition 2.1 The *approximate variation* of a function $f \in M^T$ is the one-parameter family $\{V_\varepsilon(f, T)\}_{\varepsilon > 0}$ of *ε-variations* $V_\varepsilon(f, T)$ defined, for each $\varepsilon > 0$, by

$$V_\varepsilon(f, T) = \inf\{V(g, T) : g \in \mathrm{BV}(T; M) \text{ and } d_{\infty, T}(f, g) \le \varepsilon\} \qquad (2.6)$$

(with the convention that $\inf \varnothing = \infty$).

The notion of ε-variation, which plays a crucial role in this work, is originally due to Fraňková [43, Definition 3.2] for $T = I = [a, b]$ and $M = \mathbb{R}^N$. It was also considered and extended in [29, Sections 4, 6] to any $T \subset \mathbb{R}$ and metric space (M, d), and [30] for metric space valued functions of two variables.

A few comments concerning Definition 2.1 are in order. It is efficient to rewrite (2.6) as

$$V_\varepsilon(f, T) = \inf\{V(g, T) : g \in G_{\varepsilon, T}(f)\},$$

where

$$G_{\varepsilon, T}(f) = \{g \in \mathrm{BV}(T; M) : d_{\infty, T}(f, g) \le \varepsilon\}$$

is the intersection of $\mathrm{BV}(T; M)$ and the closed ball (in the uniform metric) of radius ε centered at $f \in M^T$. So, we obtain the value $V_\varepsilon(f, T)$ if we "minimize" the lower semicontinuous functional $g \mapsto V(g, T)$ over the metric subspace $G_{\varepsilon, T}(f)$ of $\mathrm{BV}(T; M)$. Clearly, $V_\varepsilon(f, T) \in [0, \infty]$, and the value $V_\varepsilon(f, T)$ does not change if we replace condition $g \in \mathrm{BV}(T; M)$ at the right-hand side of (2.6) by less restrictive conditions $g \in M^T$ or $g \in \mathrm{B}(T; M)$.

Condition $V_\varepsilon(f, T) = \infty$ simply means that $G_{\varepsilon, T}(f) = \varnothing$, i.e.,

$$V(g, T) = \infty \text{ for all } g \in M^T \text{ such that } d_{\infty, T}(f, g) \le \varepsilon. \qquad (2.7)$$

The finiteness of $V_\varepsilon(f, T)$ is equivalent to the following: for any number η such that $\eta > V_\varepsilon(f, T)$ there is a function $g \in \mathrm{BV}(T; M)$, depending on ε and η, such that $d_{\infty, T}(f, g) \le \varepsilon$ and $V_\varepsilon(f, T) \le V(g, T) \le \eta$. Given a positive integer $k \in \mathbb{N}$, setting $\eta = V_\varepsilon(f, T) + (1/k)$, we find that there is $g_k^\varepsilon \in \mathrm{BV}(T; M)$ such that

$$d_{\infty, T}(f, g_k^\varepsilon) \le \varepsilon \quad \text{and} \quad V_\varepsilon(f, T) \le V(g_k^\varepsilon, T) \le V_\varepsilon(f, T) + (1/k); \qquad (2.8)$$

in particular, (2.8) implies $V_\varepsilon(f, T) = \lim_{k \to \infty} V(g_k^\varepsilon, T)$.

Given $\varepsilon > 0$, condition $V_\varepsilon(f, T) = 0$ is characterized as follows (cf. (2.8)):

$$\exists \{g_k\} \subset \mathrm{BV}(T; M) \text{ such that } \sup_{k \in \mathbb{N}} d_{\infty, T}(f, g_k) \le \varepsilon \text{ and } \lim_{k \to \infty} V(g_k, T) = 0.$$
$$(2.9)$$

In particular, if $g_k = c$ is a constant function on T for all $k \in \mathbb{N}$, we have:

$$\text{if } d_{\infty,T}(f, c) \leq \varepsilon, \text{ then } V_\varepsilon(f, T) = 0. \tag{2.10}$$

This is the case when $|f(T)| \leq \varepsilon$; more explicitly, (2.10) implies

$$\text{if } \varepsilon > 0 \text{ and } |f(T)| \leq \varepsilon, \text{ then } V_\varepsilon(f, T) = 0. \tag{2.11}$$

In fact, fixing $t_0 \in T$, we may define a constant function by $c(t) = f(t_0)$ for all $t \in T$, so that $d_{\infty,T}(f, c) \leq |f(T)| \leq \varepsilon$.

The lower bound $|f(T)|$ for ε in (2.11) can be refined provided $f \in M^T$ satisfies certain additional assumptions. By (2.3), $|f(T)| \leq 2d_{\infty,T}(f, c)$ for every constant function $c \in M^T$. Now, if $|f(T)| = 2d_{\infty,T}(f, c)$ for some c, we have:

$$\text{if } \varepsilon > 0 \text{ and } \varepsilon \geq |f(T)|/2, \text{ then } V_\varepsilon(f, T) = 0. \tag{2.12}$$

To see this, note that $d_{\infty,T}(f, c) = |f(T)|/2 \leq \varepsilon$ and apply (2.10).

The number $|f(T)|/2$ in (2.12) is the *best possible* lower bound for ε, for which we may have $V_\varepsilon(f, T) = 0$; in fact, by Lemma 2.5(b) (see below), if $V_\varepsilon(f, T) = 0$, then $|f(T)| \leq 2\varepsilon$, i.e., $\varepsilon \geq |f(T)|/2$. In other words, if $0 < \varepsilon < |f(T)|/2$, then $V_\varepsilon(f, T) \neq 0$.

To present an example of condition $|f(T)| = 2d_{\infty,T}(f, c)$, suppose $f \in M^T$ has only two values, i.e., $f(T) = \{x, y\}$ for some $x, y \in M, x \neq y$. Then, the mentioned condition is of the form

$$d(x, y) = 2 \max\{d(x, c), d(y, c)\} \quad \text{for some} \quad c \in M. \tag{2.13}$$

Condition (2.13) is satisfied for such f if, for instance, $(M, \|\cdot\|)$ is a *normed linear space* over $\mathbb{K} = \mathbb{R}$ or $\mathbb{C}$ (always equipped) with the *induced metric* $d(u, v) = \|u - v\|, u, v \in M$. In fact, we may set $c(t) = c = (x + y)/2, t \in T$. Note that (2.13) is concerned with a certain form of "convexity" of metric space (M, d) (cf. [29, Example 1]).

If $f(T) = \{x, y, z\}$, condition $|f(T)| = 2d_{\infty,T}(f, c)$ is of the form

$$\max\{d(x, y), d(x, z), d(y, z)\} = 2 \max\{d(x, c), d(y, c), d(z, c)\}.$$

Some elementary properties of ε-variation(s) of $f \in M^T$ are gathered in

Lemma 2.2

(a) *The function* $\varepsilon \mapsto V_\varepsilon(f, T) : (0, \infty) \to [0, \infty]$ *is nonincreasing, and so, the following inequalities hold (for one-sided limits):*

$$V_{\varepsilon+0}(f, T) \leq V_\varepsilon(f, T) \leq V_{\varepsilon-0}(f, T) \text{ in } [0, \infty] \text{ for all } \varepsilon > 0. \tag{2.14}$$

(b) *If* $\emptyset \neq T_1 \subset T_2 \subset T$, *then* $V_\varepsilon(f, T_1) \leq V_\varepsilon(f, T_2)$ *for all* $\varepsilon > 0$.

Proof

(a) Let $0 < \varepsilon_1 < \varepsilon_2$. Since $d_{\infty,T}(f, g) \leq \varepsilon_1$ implies $d_{\infty,T}(f, g) \leq \varepsilon_2$ for $g \in M^T$, we get $G_{\varepsilon_1,T}(f) \subset G_{\varepsilon_2,T}(f)$, and so, by (2.6), $V_{\varepsilon_2}(f, T) \leq V_{\varepsilon_1}(f, T)$.

(b) Given $g \in M^T$, $T_1 \subset T_2$ implies $d_{\infty,T_1}(f, g) \leq d_{\infty,T_2}(f, g)$. So, for any $\varepsilon > 0$, $G_{\varepsilon,T_2}(f) \subset G_{\varepsilon,T_1}(f)$, which, by (2.6), yields $V_{\varepsilon}(f, T_1) \leq V_{\varepsilon}(f, T_2)$.   □

2.3   Variants of approximate variation

Here we consider two modifications of the notion of approximate variation.

The first one is obtained if we replace the inequality $\leq \varepsilon$ in (2.6) by the strict inequality $< \varepsilon$; namely, given $f \in M^T$ and $\varepsilon > 0$, we set

$$V_{\varepsilon}'(f, T) = \inf\{V(g, T) : g \in \mathrm{BV}(T; M) \text{ such that } d_{\infty,T}(f, g) < \varepsilon\} \qquad (2.15)$$

($\inf \varnothing = \infty$). Clearly, Lemma 2.2 holds for $V_{\varepsilon}'(f, T)$. More specific properties of $V_{\varepsilon}'(f, T)$ are established in the following.

Proposition 2.3 *Given $f \in M^T$, we have:*

(a) *The function $\varepsilon \mapsto V_{\varepsilon}'(f, T) : (0, \infty) \to [0, \infty]$ is continuous from the left on $(0, \infty)$;*

(b) *$V_{\varepsilon_1}'(f, T) \leq V_{\varepsilon+0}'(f, T) \leq V_{\varepsilon}(f, T) \leq V_{\varepsilon-0}(f, T) \leq V_{\varepsilon}'(f, T)$ for all $0 < \varepsilon < \varepsilon_1$.*

Proof

(a) In view of (2.14) for $V_{\varepsilon}'(f, T)$, given $\varepsilon > 0$, it suffices to show that $V_{\varepsilon-0}'(f, T) \leq V_{\varepsilon}'(f, T)$ provided $V_{\varepsilon}'(f, T) < \infty$. By definition (2.15), for any number $\eta > V_{\varepsilon}'(f, T)$ there is $g = g_{\varepsilon,\eta} \in \mathrm{BV}(T; M)$ such that $d_{\infty,T}(f, g) < \varepsilon$ and $V(g, T) \leq \eta$. If a number ε' is such that $d_{\infty,T}(f, g) < \varepsilon' < \varepsilon$, then (2.15) implies $V_{\varepsilon'}'(f, T) \leq V(g, T) \leq \eta$. Passing to the limit as $\varepsilon' \to \varepsilon - 0$, we get $V_{\varepsilon-0}'(f, T) \leq \eta$ for all $\eta > V_{\varepsilon}'(f, T)$, and so, $V_{\varepsilon-0}'(f, T) \leq V_{\varepsilon}'(f, T)$.

(b) To prove the first inequality, we note that $V_{\varepsilon_1}'(f, T) \leq V_{\varepsilon'}'(f, T)$ for all ε' with $\varepsilon < \varepsilon' < \varepsilon_1$. It remains to pass to the limit as $\varepsilon' \to \varepsilon + 0$.

For the second inequality, let $g \in G_{\varepsilon,T}(f)$, i.e., $g \in \mathrm{BV}(T; M)$ and $d_{\infty,T}(f, g) \leq \varepsilon$. Then, for any number ε' such that $\varepsilon < \varepsilon'$, by virtue of (2.15), $V_{\varepsilon'}'(f, T) \leq V(g, T)$, and so, as $\varepsilon' \to \varepsilon + 0$, $V_{\varepsilon+0}'(f, T) \leq V(g, T)$. Taking the infimum over all functions $g \in G_{\varepsilon,T}(f)$, we obtain the second inequality.

The third inequality is a consequence of (2.14).

Since the set $\{g \in \mathrm{BV}(T; M) : d_{\infty,T}(f, g) < \varepsilon\}$ is contained in $G_{\varepsilon,T}(f)$, we have $V_{\varepsilon}(f, T) \leq V_{\varepsilon}'(f, T)$. Replacing ε by ε' with $0 < \varepsilon' < \varepsilon$, we get $V_{\varepsilon'}(f, T) \leq V_{\varepsilon'}'(f, T)$, and so, passing to the limit as $\varepsilon' \to \varepsilon - 0$ and taking into account item (a) above, we arrive at the fourth inequality.   □

In contrast to Proposition 2.3(a), it will be shown in Lemma 2.14(a) that the function $\varepsilon \mapsto V_\varepsilon(f, T)$ is continuous from the right on $(0, \infty)$ *only* under the additional assumption on the metric space (M, d) (to be *proper*).

In the case when $T = I = [a, b]$, the second variant of the approximate variation is obtained if we replace the set of functions of bounded variation $\mathrm{BV}(I; M)$ in (2.6) by the set of step functions $\mathrm{St}(I; M)$: given $f \in M^I$, we put

$$V_\varepsilon^s(f, I) = \inf\{V(g, I) : g \in \mathrm{St}(I; M) \text{ and } d_{\infty,I}(f, g) \le \varepsilon\}, \quad \varepsilon > 0 \qquad (2.16)$$

$(\inf \varnothing = \infty)$. Clearly, $V_\varepsilon^s(f, I)$ has the properties from Lemma 2.2.

Proposition 2.4 $V_{\varepsilon+0}^s(f, I) \le V_\varepsilon(f, I) \le V_\varepsilon^s(f, I)$ *for all* $f \in M^I$ *and* $\varepsilon > 0$.

Proof By (2.4), $\{g \in \mathrm{St}(I; M) : d_{\infty,I}(f, g) \le \varepsilon\} \subset G_{\varepsilon,I}(f)$, and so, (2.6) and (2.16) imply the right-hand side inequality.

In order to prove the left-hand side inequality, we may assume that $V_\varepsilon(f, I) < \infty$. By (2.6), for any $\eta > V_\varepsilon(f, I)$ there is $g = g_{\varepsilon,\eta} \in \mathrm{BV}(I; M)$ such that $d_{\infty,I}(f, g) \le \varepsilon$ and $V(g, I) \le \eta$. Since $g \in \mathrm{BV}(I; M) \subset \mathrm{Reg}(I; M)$, by virtue of (2.5), there is a sequence $\{g_j\} \subset \mathrm{St}(I; M)$ such that $g_j \rightrightarrows g$ on I and $V(g_j, I) \le V(g, I)$ for all natural j. Hence $\limsup_{j\to\infty} V(g_j, I) \le V(g, I)$ and, by property (V.3) (p. 8), $V(g, I) \le \liminf_{j\to\infty} V(g_j, I)$, and so, $\lim_{j\to\infty} V(g_j, I) = V(g, I)$. Now, let $\varepsilon' > 0$ be arbitrary. Then, there is $j_1 = j_1(\varepsilon') \in \mathbb{N}$ such that $V(g_j, I) \le V(g, I) + \varepsilon'$ for all $j \ge j_1$, and, since $g_j \rightrightarrows g$ on I, there is $j_2 = j_2(\varepsilon') \in \mathbb{N}$ such that $d_{\infty,I}(g_j, g) \le \varepsilon'$ for all $j \ge j_2$. Noting that, for all $j \ge \max\{j_1, j_2\}$, $g_j \in \mathrm{St}(I; M)$ and

$$d_{\infty,I}(f, g_j) \le d_{\infty,I}(f, g) + d_{\infty,I}(g, g_j) \le \varepsilon + \varepsilon',$$

by definition (2.16) of $V_\varepsilon^s(f, I)$, we get

$$V_{\varepsilon+\varepsilon'}^s(f, I) \le V(g_j, I) \le V(g, I) + \varepsilon' \le \eta + \varepsilon'.$$

Passing to the limit as $\varepsilon' \to +0$, we find $V_{\varepsilon+0}^s(f, I) \le \eta$ for all $\eta > V_\varepsilon(f, I)$, and so, $V_{\varepsilon+0}^s(f, I) \le V_\varepsilon(f, I)$. $\qquad\square$

Propositions 2.3(b) and 2.4 show that the quantities $V_\varepsilon'(f, T)$ and $V_\varepsilon^s(f, I)$ are somehow "equivalent" to $V_\varepsilon(f, T)$, so their theories will no longer be developed in the sequel, and the theory of $V_\varepsilon(f, T)$ is sufficient for our purposes.

2.4 Properties of approximate variation

In order to effectively calculate the approximate variation of a function, we need more of its properties. Item (a) in the next lemma justifies the term "approximate variation," introduced in Definition 2.1.

Lemma 2.5 *Given $f \in M^T$, we have:*

(a) $\lim_{\varepsilon \to +0} V_\varepsilon(f, T) = \sup_{\varepsilon > 0} V_\varepsilon(f, T) = V(f, T)$;

(b) $|f(T)| \le V_\varepsilon(f, T) + 2\varepsilon$ *for all* $\varepsilon > 0$;

(c) $|f(T)| = \infty$ *(i.e., $f \notin B(T; M)$) if and only if* $V_\varepsilon(f, T) = \infty$ *for all* $\varepsilon > 0$;

(d) $\inf_{\varepsilon > 0}(V_\varepsilon(f, T) + \varepsilon) \le |f(T)| \le \inf_{\varepsilon > 0}(V_\varepsilon(f, T) + 2\varepsilon)$;

(e) $|f(T)| = 0$ *(i.e., f is constant) if and only if* $V_\varepsilon(f, T) = 0$ *for all* $\varepsilon > 0$;

(f) *if* $0 < \varepsilon < |f(T)|$, *then* $\max\{0, |f(T)| - 2\varepsilon\} \le V_\varepsilon(f, T) \le V(f, T)$.

Proof

(a) By Lemma 2.2(a), $C \equiv \lim_{\varepsilon \to +0} V_\varepsilon(f, T) = \sup_{\varepsilon > 0} V_\varepsilon(f, T)$ is well-defined in $[0, \infty]$. First, we assume that $f \in BV(T; M)$. Since $f \in G_{\varepsilon, T}(f)$ for all $\varepsilon > 0$, definition (2.6) implies $V_\varepsilon(f, T) \le V(f, T)$ for all $\varepsilon > 0$, and so, $C \le V(f, T) < \infty$. Now, we prove that $V(f, T) \le C$. By definition of C, for every $\eta > 0$, there is $\delta = \delta(\eta) > 0$ such that $V_\varepsilon(f, T) < C + \eta$ for all $\varepsilon \in (0, \delta)$. Let $\{\varepsilon_k\}_{k=1}^\infty \subset (0, \delta)$ be such that $\varepsilon_k \to 0$ as $k \to \infty$. For every $k \in \mathbb{N}$, the definition of $V_{\varepsilon_k}(f, T) < C + \eta$ implies the existence of $g_k \in BV(T; M)$ such that $d_{\infty, T}(f, g_k) \le \varepsilon_k$ and $V(g_k, T) \le C + \eta$. Since $\varepsilon_k \to 0$, $g_k \rightrightarrows f$ on T, and so, property **(V.3)** on p. 9 yields

$$V(f, T) \le \liminf_{k \to \infty} V(g_k, T) \le C + \eta \quad \text{for all} \quad \eta > 0,$$

whence $V(f, T) \le C < \infty$. Thus, C and $V(f, T)$ are finite or not simultaneously, and $C = V(f, T)$, which establishes (a).

(b) The inequality is clear if $V_\varepsilon(f, T) = \infty$, so we assume that $V_\varepsilon(f, T)$ is finite. By definition (2.6), for every $\eta > V_\varepsilon(f, T)$ there is $g = g_\eta \in BV(T; M)$ such that $d_{\infty, T}(f, g) \le \varepsilon$ and $V(g, T) \le \eta$. Inequality (2.3) and property **(V.1)** on p. 8 imply

$$|f(T)| \le |g(T)| + 2d_{\infty, T}(f, g) \le V(g, T) + 2\varepsilon \le \eta + 2\varepsilon.$$

It remains to take into account the arbitrariness of $\eta > V_\varepsilon(f, T)$.

(c) The necessity is a consequence of item (b). To prove the sufficiency, assume, on the contrary, that $|f(T)| < \infty$. Then, by (2.11), for any $\varepsilon > |f(T)|$, we have $V_\varepsilon(f, T) = 0$, which contradicts the assumption $V_\varepsilon(f, T) = \infty$.

(d) The right-hand side inequality is equivalent to item (b). To establish the left-hand side inequality, we note that if $|f(T)| < \infty$ and $\varepsilon > |f(T)|$, then, by (2.11), $V_\varepsilon(f, T) = 0$, and so,

$$|f(T)| = \inf_{\varepsilon > |f(T)|} \varepsilon = \inf_{\varepsilon > |f(T)|} (V_\varepsilon(f, T) + \varepsilon) \ge \inf_{\varepsilon > 0}(V_\varepsilon(f, T) + \varepsilon).$$

Now, if $|f(T)| = \infty$, then, by item (c), $V_\varepsilon(f, T) + \varepsilon = \infty$ for all $\varepsilon > 0$, and so, $\inf_{\varepsilon > 0}(V_\varepsilon(f, T) + \varepsilon) = \infty$.

(e) ($\Rightarrow$) Since f is constant on T, $f \in BV(T; M)$ and $d_{\infty,T}(f, f) = 0 < \varepsilon$, and
so, definition (2.6) implies $0 \le V_\varepsilon(f, T) \le V(f, T) = 0$.
    ($\Leftarrow$) By virtue of item (d), if $V_\varepsilon(f, T) = 0$ for all $\varepsilon > 0$, then $|f(T)| = 0$.

(f) We may assume that $f \in B(T; M)$. By item (a), $V_\varepsilon(f, T) \le V(f, T)$, and by
item (b), $|f(T)| - 2\varepsilon \le V_\varepsilon(f, T)$ for all $0 < \varepsilon < |f(T)|/2$. It is also clear that,
for all $|f(T)|/2 \le \varepsilon < |f(T)|$, we have $0 \le V_\varepsilon(f, T)$. $\square$

Remark 2.6 By (2.11) and Lemma 2.5(c), (e), the ε-variation $V_\varepsilon(f, T)$, initially
defined for all $\varepsilon > 0$ and $f \in M^T$, is *completely characterized* whenever $\varepsilon > 0$ and
$f \in B(T; M)$ are such that $0 < \varepsilon < |f(T)|$.

    The sharpness (=exactness) of assertions in Lemma 2.5(b), (d) is presented in
Example 3.6(b–d) on pp. 35–37 (for (a), (b), (f), see Example 2.7).

    Now we consider a simple example of the explicit evaluation of the approximate
variation (which later on will be generalized, cf. Example 3.1).

Example 2.7 Let $f : T = [0, 1] \to [0, 1]$ be given by $f(t) = t$. We are going
to evaluate $V_\varepsilon(f, T)$, $\varepsilon > 0$. Since $|f(T)| = 1$, by (2.11), $V_\varepsilon(f, T) = 0$ for all
$\varepsilon \ge 1 = |f(T)|$. Moreover, if $c(t) \equiv 1/2$ on T, then $|f(t) - c(t)| \le 1/2$ for
all $t \in T$, and so, $V_\varepsilon(f, T) = 0$ for all $\varepsilon \ge 1/2$. Now, suppose $0 < \varepsilon < 1/2$.
By Lemma 2.5(f), $V_\varepsilon(f, T) \ge 1 - 2\varepsilon$. To establish the reverse inequality, define
$g : T \to \mathbb{R}$ by

$$g(t) = (1 - 2\varepsilon)t + \varepsilon, \quad 0 \le t \le 1$$

(draw the graph on the plane). Clearly, g is increasing on $[0, 1]$ and, for all $t \in [0, 1]$,

$$f(t) - \varepsilon = t - \varepsilon = t - 2\varepsilon + \varepsilon \le g(t) = t - 2\varepsilon t + \varepsilon \le t + \varepsilon = f(t) + \varepsilon,$$

i.e., $d_{\infty,T}(f, g) = \sup_{t \in T} |f(t) - g(t)| \le \varepsilon$. Since g is increasing on $[0, 1]$,

$$V(g, T) = g(1) - g(0) = (1 - 2\varepsilon + \varepsilon) - \varepsilon = 1 - 2\varepsilon,$$

and so, by definition (2.6), we get $V_\varepsilon(f, T) \le V(g, T) = 1 - 2\varepsilon$. Thus,

$$\text{if } f(t) = t, \text{ then } V_\varepsilon(f, [0, 1]) = \begin{cases} 1 - 2\varepsilon & \text{if } 0 < \varepsilon < 1/2, \\ 0 & \text{if } \varepsilon \ge 1/2. \end{cases}$$

Lemma 2.8 (Semi-additivity of ε-variation) *Given $f \in M^T$, $\varepsilon > 0$, and $t \in T$,
if $T_1 = T \cap (-\infty, t]$ and $T_2 = T \cap [t, \infty)$, then we have:*

$$V_\varepsilon(f, T_1) + V_\varepsilon(f, T_2) \le V_\varepsilon(f, T) \le V_\varepsilon(f, T_1) + V_\varepsilon(f, T_2) + 2\varepsilon.$$

Proof

1. First, we prove the left-hand side inequality. We may assume that $V_\varepsilon(f, T)$ is finite (otherwise, the inequality is obvious). Given $\eta > V_\varepsilon(f, T)$, by (2.6), there is $g = g_\eta \in BV(T; M)$ such that $d_{\infty,T}(f, g) \leq \varepsilon$ and $V(g, T) \leq \eta$. We set $g_1(s) = g(s)$ for all $s \in T_1$ and $g_2(s) = g(s)$ for all $s \in T_2$, and note that $g_1(t) = g(t) = g_2(t)$. Since, for $i = 1, 2$, we have $d_{\infty,T_i}(f, g_i) \leq d_{\infty,T}(f, g) \leq \varepsilon$ and $g_i \in BV(T_i; M)$, by (2.6), we find $V_\varepsilon(f, T_i) \leq V(g_i, T_i)$, and so, the additivity property (**V.2**) of V (on p. 8) implies

$$V_\varepsilon(f, T_1) + V_\varepsilon(f, T_2) \leq V(g_1, T_1) + V(g_2, T_2) = V(g, T_1) + V(g, T_2)$$

$$= V(g, T) \leq \eta \quad \text{for all} \quad \eta > V_\varepsilon(f, T).$$

This establishes the left-hand side inequality.

2. Now, we prove the right-hand side inequality. We may assume that $V_\varepsilon(f, T_1)$ and $V_\varepsilon(f, T_2)$ are finite (otherwise, our inequality becomes $\infty = \infty$). We may also assume that $T \cap (-\infty, t) \neq \varnothing$ and $T \cap (t, \infty) \neq \varnothing$ (for, otherwise, we have $T_1 = T \cap (-\infty, t] = \{t\}$ and $T_2 = T$, or $T_2 = T \cap [t, \infty) = \{t\}$ and $T_1 = T$, respectively, and the inequality is clear). By definition (2.6), for $i = 1, 2$, given $\eta_i > V_\varepsilon(f, T_i)$, there exists $g_i \in BV(T_i; M)$ such that $d_{\infty,T_i}(f, g_i) \leq \varepsilon$ and $V(g_i, T_i) \leq \eta_i$. Given $u \in M$ (to be specified below), we define $g \in BV(T; M)$ by

$$g(s) = \begin{cases} g_1(s) & \text{if } s \in T \cap (-\infty, t), \\ u & \text{if } s = t, \\ g_2(s) & \text{if } s \in T \cap (t, \infty). \end{cases}$$

Arguing with partitions of T_i for $i = 1, 2$ (see step 3 below) and applying the triangle inequality for d, we get

$$V(g, T_i) \leq V(g_i, T_i) + d(g(t), g_i(t)) \leq \eta_i + d(u, g_i(t)). \tag{2.17}$$

By the additivity (**V.2**) of the functional V, we find

$$V(g, T) = V(g, T_1) + V(g, T_2) \leq \eta_1 + d(u, g_1(t)) + \eta_2 + d(u, g_2(t)). \tag{2.18}$$

Now, we set $u = g_1(t)$ (by symmetry, we may set $u = g_2(t)$ as well). Since $g = g_1$ on $T_1 = T \cap (-\infty, t]$ and $g = g_2$ on $T \cap (t, \infty) \subset T_2$, we get

$$d_{\infty,T}(f, g) \leq \max\{d_{\infty,T_1}(f, g_1), d_{\infty,T_2}(f, g_2)\} \leq \varepsilon. \tag{2.19}$$

Noting that (cf. (2.18))

$$d(u, g_2(t)) = d(g_1(t), g_2(t)) \le d(g_1(t), f(t)) + d(f(t), g_2(t))$$

$$\le d_{\infty, T_1}(g_1, f) + d_{\infty, T_2}(f, g_2) \le \varepsilon + \varepsilon = 2\varepsilon,$$

we conclude from (2.6), (2.19), and (2.18) that

$$V_\varepsilon(f, T) \le V(g, T) \le \eta_1 + \eta_2 + 2\varepsilon.$$

The arbitrariness of numbers $\eta_1 > V_\varepsilon(f, T_1)$ and $\eta_2 > V_\varepsilon(f, T_2)$ proves the desired inequality.

3. *Proof of (2.17)* for $i = 1$ (the case $i = 2$ is similar). Let $\{t_k\}_{k=0}^m \subset T_1$ be a partition of T_1, i.e., $t_0 < t_1 < \cdots < t_{m-1} < t_m = t$. Since $g(s) = g_1(s)$ for $s \in T$, $s < t$, we have:

$$\sum_{k=1}^m d(g(t_k), g(t_{k-1})) = \sum_{k=1}^{m-1} d(g(t_k), g(t_{k-1})) + d(g(t_m), g(t_{m-1}))$$

$$= \sum_{k=1}^{m-1} d(g_1(t_k), g_1(t_{k-1})) + d(g_1(t_m), g_1(t_{m-1}))$$

$$+ d(g(t_m), g(t_{m-1})) - d(g_1(t_m), g_1(t_{m-1}))$$

$$\le V(g_1, T_1) + |d(g(t), g_1(t_{m-1})) - d(g_1(t), g_1(t_{m-1}))|$$

$$\le V(g_1, T_1) + d(g(t), g_1(t)),$$

where the last inequality is due to the triangle inequality for d. Taking the supremum over all partitions of T_1, we obtain the left-hand side inequality in (2.17) for $i = 1$.   □

Remark 2.9 The informative part of Lemma 2.8 concerns the case when $f \in B(T; M)$ and $0 < \varepsilon < |f(T)|$; if fact, if $\varepsilon \ge |f(T)|$, then $\varepsilon \ge |f(T_1)|$ and $\varepsilon \ge |f(T_2)|$, and so, by (2.11), $V_\varepsilon(f, T) = V_\varepsilon(f, T_1) = V_\varepsilon(f, T_2) = 0$. The sharpness of the inequalities in Lemma 2.8 is shown in Example 3.2.

Interestingly, the approximate variation characterizes regulated functions. The following lemma is Fraňková's result [43, Proposition 3.4] rewritten from the case when $I = [a, b]$ and $M = \mathbb{R}^N$ to the case of an arbitrary metric space (M, d) (which was announced in [29, equality (4.2)]).

Lemma 2.10 $Reg(I; M) = \{f \in M^I : V_\varepsilon(f, I) < \infty \text{ for all } \varepsilon > 0.\}$

Proof ($\subset$) If $f \in \text{Reg}(I; M)$, then, by (2.5), given $\varepsilon > 0$, there is $g_\varepsilon \in \text{St}(I; M)$ such that $d_{\infty,I}(f, g_\varepsilon) \le \varepsilon$. Since $g_\varepsilon \in \text{BV}(I; M)$, (2.6) implies $V_\varepsilon(f, I) \le V(g_\varepsilon, I) < \infty$.

($\supset$) Suppose $f \in M^I$ and $V_\varepsilon(f, I) < \infty$ for all $\varepsilon > 0$. Given $a < \tau \le b$, let us show that $d(f(s), f(t)) \to 0$ as $s, t \to \tau - 0$ (the arguments for $a \le \tau' < b$ and the limit as $s, t \to \tau' + 0$ are similar). Let $\varepsilon > 0$ be arbitrary. We define the ε-*variation function* by $\varphi_\varepsilon(t) = V_\varepsilon(f, [a, t]), t \in I$. By Lemma 2.2(b), $0 \le \varphi_\varepsilon(s) \le \varphi_\varepsilon(t) \le V_\varepsilon(f, I) < \infty$ for all $s, t \in I, s \le t$, i.e., $\varphi_\varepsilon : I \to [0, \infty)$ is bounded and nondecreasing, and so, the left limit $\lim_{t \to \tau - 0} \varphi_\varepsilon(t)$ exists in $[0, \infty)$. Hence, there is $\delta = \delta(\varepsilon) \in (0, \tau - a]$ such that $|\varphi_\varepsilon(t) - \varphi_\varepsilon(s)| < \varepsilon$ for all $s, t \in [\tau - \delta, \tau)$. Now, let $s, t \in [\tau - \delta, \tau), s \le t$, be arbitrary. Lemma 2.8 (with $T_1 = [a, s], T_2 = [s, t]$, and $T = [a, t]$) implies

$$V_\varepsilon(f, [s, t]) \le \varphi_\varepsilon(t) - \varphi_\varepsilon(s) < \varepsilon.$$

By the definition of $V_\varepsilon(f, [s, t])$, there is $g = g_\varepsilon \in \text{BV}([s, t]; M)$ such that

$$d_{\infty,[s,t]}(f, g) \le \varepsilon \quad \text{and} \quad V(g, [s, t]) \le \varepsilon.$$

Thus, by virtue of (2.2),

$$d(f(s), f(t)) \le d(g(s), g(t)) + 2d_{\infty,[s,t]}(f, g) \le V(g, [s, t]) + 2\varepsilon \le 3\varepsilon.$$

This completes the proof of the equality $\lim_{I \ni s,t \to \tau - 0} d(f(s), f(t)) = 0$. $\qquad\square$

Remark 2.11 We presented a direct proof of assertion ($\supset$) in Lemma 2.10. Indirectly, we may argue as in [43, Proposition 3.4] as follows. Since, for each $k \in \mathbb{N}$, $V_{1/k}(f, I) < \infty$, by definition (2.6), there is $g_k \in \text{BV}(I; M)$ such that $d_{\infty,I}(f, g_k) \le 1/k$ (and $V(g_k, I) \le V_{1/k}(f, I) + (1/k)$). Noting that $g_k \rightrightarrows f$ on I, each $g_k \in \text{Reg}(I; M)$, and $\text{Reg}(I; M)$ is closed with respect to the uniform convergence, we infer that $f \in \text{Reg}(I; M)$. An illustration of Lemma 2.10 is presented in Examples 3.1 and 3.6.

Now we study the approximate variation in its interplay with the uniform convergence of sequences of functions (see also Examples 3.7–3.9).

Lemma 2.12 *Suppose* $f \in M^T$, $\{f_j\} \subset M^T$, *and* $f_j \rightrightarrows f$ *on* T. *We have:*

(a) $V_{\varepsilon+0}(f, T) \le \liminf_{j \to \infty} V_\varepsilon(f_j, T) \le \limsup_{j \to \infty} V_\varepsilon(f_j, T) \le V_{\varepsilon-0}(f, T)$ *for all*
    $\varepsilon > 0$;
(b) *If* $V_\varepsilon(f_j, T) < \infty$ *for all* $\varepsilon > 0$ *and* $j \in \mathbb{N}$, *then* $V_\varepsilon(f, T) < \infty$ *for all* $\varepsilon > 0$.

Proof

(a) Only the first and the last inequalities are to be verified.

1. In order to prove the first inequality, we may assume (passing to a suitable subsequence of $\{f_j\}$ if necessary) that the right-hand side (i.e., the lim inf) is equal to $C \equiv \lim_{j \to \infty} V_\varepsilon(f_j, T) < \infty$. Suppose $\eta > 0$ is given arbitrarily. Then, there is $j_0 = j_0(\eta) \in \mathbb{N}$ such that $V_\varepsilon(f_j, T) \le C + \eta$ for all $j \ge j_0$. By the definition of $V_\varepsilon(f_j, T)$, for every $j \ge j_0$ there is $g_j = g_{j,\eta} \in BV(T; M)$ such that $d_{\infty,T}(f_j, g_j) \le \varepsilon$ and $V(g_j, T) \le V_\varepsilon(f_j, T) + \eta$. Since $f_j \rightrightarrows f$ on T, we have $d_{\infty,T}(f_j, f) \to 0$ as $j \to \infty$, and so, there is $j_1 = j_1(\eta) \in \mathbb{N}$ such that $d_{\infty,T}(f_j, f) \le \eta$ for all $j \ge j_1$. Noting that

$$d_{\infty,T}(f, g_j) \le d_{\infty,T}(f, f_j) + d_{\infty,T}(f_j, g_j) \le \eta + \varepsilon \quad \text{for all } j \ge \max\{j_0, j_1\},$$

we find, by virtue of definition (2.6),

$$V_{\eta+\varepsilon}(f, T) \le V(g_j, T) \le V_\varepsilon(f_j, T) + \eta \le (C + \eta) + \eta = C + 2\eta.$$

Passing to the limit as $\eta \to +0$, we arrive at $V_{\varepsilon+0}(f, T) \le C$, which was to be proved.

2. To establish the last inequality, with no loss of generality we may assume that $V_{\varepsilon-0}(f, T) < \infty$. Given $\eta > 0$, there is $\delta = \delta(\eta, \varepsilon) \in (0, \varepsilon)$ such that if $\varepsilon' \in [\varepsilon - \delta, \varepsilon)$, we have $V_{\varepsilon'}(f, T) \le V_{\varepsilon-0}(f, T) + \eta$. Since $f_j \rightrightarrows f$ on T, given $\varepsilon - \delta \le \varepsilon' < \varepsilon$, there is $j_0 = j_0(\varepsilon', \varepsilon) \in \mathbb{N}$ such that $d_{\infty,T}(f_j, f) \le \varepsilon - \varepsilon'$ for all $j \ge j_0$. By the definition of $V_{\varepsilon'}(f, T)$, for every $j \in \mathbb{N}$ we find $g_j = g_{j,\varepsilon'} \in BV(T; M)$ such that $d_{\infty,T}(f, g_j) \le \varepsilon'$ and

$$V_{\varepsilon'}(f, T) \le V(g_j, T) \le V_{\varepsilon'}(f, T) + (1/j),$$

which implies $\lim_{j \to \infty} V(g_j, T) = V_{\varepsilon'}(f, T)$. Noting that, for all $j \ge j_0$,

$$d_{\infty,T}(f_j, g_j) \le d_{\infty,T}(f_j, f) + d_{\infty,T}(f, g_j) \le (\varepsilon - \varepsilon') + \varepsilon' = \varepsilon,$$

we find from (2.6) that $V_\varepsilon(f_j, T) \le V(g_j, T)$ for all $j \ge j_0$. It follows that

$$\limsup_{j \to \infty} V_\varepsilon(f_j, T) \le \lim_{j \to \infty} V(g_j, T) = V_{\varepsilon'}(f, T) \le V_{\varepsilon-0}(f, T) + \eta.$$

It remains to take into account the arbitrariness of $\eta > 0$.

(b) Let $\varepsilon > 0$ and $0 < \varepsilon' < \varepsilon$. Given $j \in \mathbb{N}$, since $V_{\varepsilon'}(f_j, T) < \infty$, by definition (2.6), there is $g_j \in BV(T; M)$ such that $d_{\infty,T}(f_j, g_j) \le \varepsilon'$ and $V(g_j, T) \le V_{\varepsilon'}(f_j, T) + 1$. Since $f_j \rightrightarrows f$ on T, there is $j_0 = j_0(\varepsilon - \varepsilon') \in \mathbb{N}$ such that $d_{\infty,T}(f_{j_0}, f) \le \varepsilon - \varepsilon'$. Noting that

$$d_{\infty,T}(f, g_{j_0}) \le d_{\infty,T}(f, f_{j_0}) + d_{\infty,T}(f_{j_0}, g_{j_0}) \le (\varepsilon - \varepsilon') + \varepsilon' = \varepsilon,$$

we get, by (2.6), $V_\varepsilon(f, T) \le V(g_{j_0}, T) \le V_{\varepsilon'}(f_{j_0}, T) + 1 < \infty.$ □

Lemma 2.13 (Change of variable in the approximate variation)   *Suppose $T \subset \mathbb{R}$, $\varphi : T \to \mathbb{R}$ is a strictly monotone function, and $f : \varphi(T) \to M$. Then[2]*

$$V_\varepsilon(f, \varphi(T)) = V_\varepsilon(f \circ \varphi, T) \quad \text{for all} \ \varepsilon > 0.$$

Proof We need the following "change of variable" formula for Jordan's variation (cf. [11, Theorem 2.20], [13, Proposition 2.1(V4)]): if $T \subset \mathbb{R}$, $\varphi : T \to \mathbb{R}$ is a (not necessarily strictly) *monotone* function and $g : \varphi(T) \to M$, then

$$V(g, \varphi(T)) = V(g \circ \varphi, T). \tag{2.20}$$

($\geq$) Suppose $V_\varepsilon(f, \varphi(T)) < \infty$. By definition (2.6), for every $\eta > V_\varepsilon(f, \varphi(T))$ there is $g \in \mathrm{BV}(\varphi(T); M)$ such that $d_{\infty,\varphi(T)}(f, g) \leq \varepsilon$ and $V(g, \varphi(T)) \leq \eta$. We have $g \circ \varphi \in M^T$,

$$d_{\infty,T}(f \circ \varphi, g \circ \varphi) = d_{\infty,\varphi(T)}(f, g) \leq \varepsilon, \tag{2.21}$$

and, by (2.20), $V(g \circ \varphi, T) = V(g, \varphi(T)) \leq \eta$. Thus, by (2.6) and (2.21),

$$V_\varepsilon(f \circ \varphi, T) \leq V(g \circ \varphi, T) \leq \eta \quad \text{for all} \ \eta > V_\varepsilon(f, \varphi(T)),$$

and so, $V_\varepsilon(f \circ \varphi, T) \leq V_\varepsilon(f, \varphi(T)) < \infty$.

($\leq$) Now, suppose $V_\varepsilon(f \circ \varphi, T) < \infty$. Then, for every $\eta > V_\varepsilon(f \circ \varphi, T)$ there exists $g \in \mathrm{BV}(T; M)$ such that $d_{\infty,T}(f \circ \varphi, g) \leq \varepsilon$ and $V(g, T) \leq \eta$. Denote by $\varphi^{-1} : \varphi(T) \to T$ the inverse function of φ. Clearly, φ^{-1} is strictly monotone on $\varphi(T)$ in the same sense as φ on T. Setting $g_1 = g \circ \varphi^{-1}$, we find $g_1 : \varphi(T) \to M$ and, by (2.21),

$$d_{\infty,\varphi(T)}(f, g_1) = d_{\infty,\varphi(T)}\big((f \circ \varphi) \circ \varphi^{-1}, g \circ \varphi^{-1}\big)$$

$$= d_{\infty,\varphi^{-1}(\varphi(T))}(f \circ \varphi, g) = d_{\infty,T}(f \circ \varphi, g) \leq \varepsilon.$$

Furthermore, by (2.20),

$$V(g_1, \varphi(T)) = V(g \circ \varphi^{-1}, \varphi(T)) = V(g, \varphi^{-1}(\varphi(T))) = V(g, T) \leq \eta.$$

Thus, $V_\varepsilon(f, \varphi(T)) \leq V(g_1, \varphi(T)) \leq \eta$ for all $\eta > V_\varepsilon(f \circ \varphi, T)$, which implies the inequality $V_\varepsilon(f, \varphi(T)) \leq V_\varepsilon(f \circ \varphi, T) < \infty$.   □

Lemma 2.13 will be applied in Example 3.5 (cf. Case $\alpha > 1$ on p. 33).

Under additional assumptions on the metric space (M, d), we get three more properties of the approximate variation. Recall that (M, d) is called *proper* (or has

---

[2] Here, as usual, $\varphi(T) = \{\varphi(t) : t \in T\}$ is the image of T under φ, and $f \circ \varphi$ is the composition of functions $\varphi : T \to \mathbb{R}$ and $f : \varphi(T) \to M$ given by $(f \circ \varphi)(t) = f(\varphi(t))$ for all $t \in T$.

the *Heine–Borel property*) if all *closed bounded* subsets of M are *compact*. For instance, if $(M, \|\cdot\|)$ is a *finite-dimensional* normed linear space with induced metric d (cf. p. 11), then (M, d) is a proper metric space. Note that a proper metric space is complete. In fact, if $\{x_j\}_{j=1}^{\infty}$ is a Cauchy sequence in M, then it is bounded and, since M is proper, the set $\{x_j : j \in \mathbb{N}\}$ is relatively compact in M. Hence a subsequence of $\{x_j\}_{j=1}^{\infty}$ converges in M to an element $x \in M$. Now, since $\{x_j\}_{j=1}^{\infty}$ is Cauchy, we get $x_j \to x$ as $j \to \infty$, which proves the completeness of M.

Lemma 2.14 *Let (M, d) be a proper metric space and $f \in M^T$. We have:*

(a) *The function $\varepsilon \mapsto V_\varepsilon(f, T)$ is continuous from the right on $(0, \infty)$;*

(b) *Given $\varepsilon > 0$, $V_\varepsilon(f, T) < \infty$ if and only if $V_\varepsilon(f, T) = V(g, T)$ for some function $g = g_\varepsilon \in G_{\varepsilon, T}(f)$ (i.e., the infimum in (2.6) is attained, and so, becomes the minimum);*

(c) *If $\{f_j\} \subset M^T$ and $f_j \to f$ on T, then $V_\varepsilon(f, T) \leq \liminf\limits_{j \to \infty} V_\varepsilon(f_j, T)$ for all $\varepsilon > 0$.*

Proof

(a) By virtue of (2.14), it suffices to show that $V_\varepsilon(f, T) \leq V_{\varepsilon+0}(f, T)$ provided $V_{\varepsilon+0}(f, T)$ is finite. In fact, given $\eta > V_{\varepsilon+0}(f, T) = \lim_{\varepsilon' \to \varepsilon+0} V_{\varepsilon'}(f, T)$, there is $\delta = \delta(\eta) > 0$ such that $\eta > V_{\varepsilon'}(f, T)$ for all ε' with $\varepsilon < \varepsilon' \leq \varepsilon + \delta$. Let $\{\varepsilon_k\}_{k=1}^{\infty}$ be a sequence such that $\varepsilon < \varepsilon_k \leq \varepsilon + \delta$ for all $k \in \mathbb{N}$ and $\varepsilon_k \to \varepsilon$ as $k \to \infty$. Given $k \in \mathbb{N}$, setting $\varepsilon' = \varepsilon_k$, we find $\eta > V_{\varepsilon_k}(f, T)$, and so, by definition (2.6), there is $g_k \in \mathrm{BV}(T; M)$ (also depending on η) such that

$$d_{\infty, T}(f, g_k) \leq \varepsilon_k \quad \text{and} \quad V(g_k, T) \leq \eta. \tag{2.22}$$

By the first inequality in (2.22), the sequence $\{g_k\}$ is pointwise bounded on T, because, given $t \in T$, by the triangle inequality for d, we have

$$d(g_k(t), g_j(t)) \leq d(g_k(t), f(t)) + d(f(t), g_j(t))$$

$$\leq d_{\infty, T}(g_k, f) + d_{\infty, T}(f, g_j) \tag{2.23}$$

$$\leq \varepsilon_k + \varepsilon_j \leq 2(\varepsilon + \delta) \quad \text{for all} \quad k, j \in \mathbb{N},$$

and since (M, d) is *proper*, the sequence $\{g_k\}$ is pointwise relatively compact on T. So, the second inequality in (2.22) and the Helly-type selection principle in $\mathrm{BV}(T; M)$ (which is property (**V.4**) on p. 9) imply the existence of a subsequence of $\{g_k\}$, again denoted by $\{g_k\}$ (and the corresponding subsequence of $\{\varepsilon_k\}$—again denoted by $\{\varepsilon_k\}$), and a function $g \in \mathrm{BV}(T; M)$ such that $g_k \to g$ pointwise on T. Noting that, by (2.22),

$$d_{\infty, T}(f, g) \leq \liminf\limits_{k \to \infty} d_{\infty, T}(f, g_k) \leq \lim\limits_{k \to \infty} \varepsilon_k = \varepsilon \tag{2.24}$$

and, by the lower semicontinuity of V (property (**V.3**) on p. 9),

$$V(g, T) \leq \liminf_{k \to \infty} V(g_k, T) \leq \eta, \tag{2.25}$$

we find, from definition (2.6), that $V_\varepsilon(f, T) \leq V(g, T) \leq \eta$. It remains to take into account the arbitrariness of $\eta > V_{\varepsilon+0}(f, T)$.

Items (b) and (c) were essentially established in [43] for $T = [a, b]$ and $M = \mathbb{R}^N$ as Propositions 3.5 and 3.6, respectively. For the sake of completeness, we present the proofs of (b) and (c) in our more general situation.

(b) The sufficiency ($\Leftarrow$) is clear. Now we establish the necessity ($\Rightarrow$). By definition (2.6), given $k \in \mathbb{N}$, there is $g_k \in \mathrm{BV}(T; M)$ such that

$$d_{\infty,T}(f, g_k) \leq \varepsilon \quad \text{and} \quad V_\varepsilon(f, T) \leq V(g_k, T) \leq V_\varepsilon(f, T) + (1/k). \tag{2.26}$$

From (2.23) and (2.26), we find $d(g_k(t), g_j(t)) \leq 2\varepsilon$ for all $k, j \in \mathbb{N}$ and $t \in T$, and so, the sequence $\{g_k\}$ is pointwise bounded on T, and since (M, d) is *proper*, $\{g_k\}$ is pointwise relatively compact on T. Moreover, by (2.26), $\sup_{k \in \mathbb{N}} V(g_k, T) \leq V_\varepsilon(f, T) + 1 < \infty$. By the Helly-type selection principle (**V.4**) in $\mathrm{BV}(T; M)$, there are a subsequence of $\{g_k\}$, again denoted by $\{g_k\}$, and a function $g \in \mathrm{BV}(T; M)$ such that $g_k \to g$ on T. As in (2.24), we get $d_{\infty,T}(f, g) \leq \varepsilon$, and so, (2.6), (2.25), and (2.26) yield

$$V_\varepsilon(f, T) \leq V(g, T) \leq \lim_{k \to \infty} V(g_k, T) = V_\varepsilon(f, T).$$

(c) Passing to a subsequence of $\{f_j\}$ (if necessary), we may assume that the right-hand side of the inequality in (c) is equal to $C_\varepsilon = \lim_{j \to \infty} V_\varepsilon(f_j, T)$ and finite. Given $\eta > C_\varepsilon$, there is $j_0 = j_0(\eta, \varepsilon) \in \mathbb{N}$ such that $\eta > V_\varepsilon(f_j, T)$ for all $j \geq j_0$. For every $j \geq j_0$, by the definition of $V_\varepsilon(f_j, T)$, there is $g_j \in \mathrm{BV}(T; M)$ such that

$$d_{\infty,T}(f_j, g_j) \leq \varepsilon \quad \text{and} \quad V(g_j, T) \leq \eta. \tag{2.27}$$

Since $f_j \to f$ pointwise on T, $\{f_j\}$ is pointwise relatively compact on T, and so, $\{f_j\}$ is pointwise bounded on T, i.e., $B(t) = \sup_{j,k \in \mathbb{N}} d(f_j(t), f_k(t)) < \infty$ for all $t \in T$. By the triangle inequality for d and (2.27), given $j, k \geq j_0$ and $t \in T$, we have

$$d(g_j(t), g_k(t)) \leq d(g_j(t), f_j(t)) + d(f_j(t), f_k(t)) + d(f_k(t), g_k(t))$$

$$\leq d_{\infty,T}(g_j, f_j) + B(t) + d_{\infty,T}(f_k, g_k) \leq B(t) + 2\varepsilon.$$

This implies that the sequence $\{g_j\}_{j=j_0}^{\infty}$ is pointwise bounded on T, and since (M, d) is *proper*, it is pointwise relatively compact on T. It follows from (2.27) that $\sup_{j \geq j_0} V(g_j, T)$ does not exceed $\eta < \infty$, and so, by the Helly-type selection principle (**V.4**) in $BV(T; M)$, there are a subsequence $\{g_{j_p}\}_{p=1}^{\infty}$ of $\{g_j\}_{j=j_0}^{\infty}$ and a function $g \in BV(T; M)$ such that $g_{j_p} \to g$ pointwise on T as $p \to \infty$. Since $f_{j_p} \to f$ pointwise on T as $p \to \infty$, we find, from (2.27) and property (**V.3**) on p. 9, that

$$d_{\infty,T}(f, g) \le \liminf_{p \to \infty} d_{\infty,T}(f_{j_p}, g_{j_p}) \le \varepsilon$$

and

$$V(g, T) \le \liminf_{p \to \infty} V(g_{j_p}, T) \le \eta.$$

Now, definition (2.6) implies $V_\varepsilon(f, T) \le V(g, T) \le \eta$ for all $\eta > C_\varepsilon$, and so, $V_\varepsilon(f, T) \le C_\varepsilon = \lim_{j \to \infty} V_\varepsilon(f_j, T)$, which was to be proved. □

Remark 2.15 The inequality in Lemma 2.14(c) agrees with the left-hand side inequality in Lemma 2.12(a): in fact, if (M, d) is *proper*, $\{f_j\} \subset M^T$, and $f_j \rightrightarrows f$ on T, then, by Lemma 2.14(a),

$$V_\varepsilon(f, T) = V_{\varepsilon+0}(f, T) \le \liminf_{j \to \infty} V_\varepsilon(f_j, T), \quad \varepsilon > 0.$$

The properness of (M, d) in Lemma 2.14 is essential: item (a) is illustrated in Example 3.6(e) on p. 37, (b)—in Example 3.10, and (c)—in Example 3.11.

Chapter 3
Examples of approximate variations

This chapter is devoted to various examples of approximate variations. In particular, it is shown that all assertions in Sect. 2.4 are sharp.

3.1 Functions with values in a normed linear space

Example 3.1 Let $T \subset \mathbb{R}$ and $(M, \| \cdot \|)$ be a normed linear space (cf. p. 11). We have $d_{\infty,T}(f, g) = \|f - g\|_{\infty,T}$, $f, g \in M^T$, where the *uniform norm* on M^T is given by

$$\|f\|_{\infty,T} = \sup_{t \in T} \|f(t)\|, \qquad f \in M^T.$$

We are going to estimate (and evaluate) the approximate variation $\{V_\varepsilon(f, T)\}_{\varepsilon>0}$ for the function $f : T \to M$ defined, for $x, y \in M$, $x \neq 0$, by

$$f(t) = \varphi(t)x + y, \; t \in T, \quad \text{where } \varphi \in \mathrm{BV}(T; \mathbb{R}) \text{ is } nonconstant. \tag{3.1}$$

To begin with, recall that $0 < |\varphi(T)| \leq V(\varphi, T) < \infty$ and

$$|\varphi(T)| = \sup_{s,t \in T} |\varphi(s) - \varphi(t)| = \mathrm{s}(\varphi) - \mathrm{i}(\varphi),$$

where $\mathrm{s}(\varphi) \equiv \mathrm{s}(\varphi, T) = \sup_{t \in T} \varphi(t)$ and $\mathrm{i}(\varphi) \equiv \mathrm{i}(\varphi, T) = \inf_{t \in T} \varphi(t)$. Moreover,

$$\left| \varphi(t) - \frac{\mathrm{i}(\varphi) + \mathrm{s}(\varphi)}{2} \right| \leq \frac{\mathrm{s}(\varphi) - \mathrm{i}(\varphi)}{2} = \frac{|\varphi(T)|}{2} \quad \text{for all } t \in T. \tag{3.2}$$

V. V. Chistyakov, *From Approximate Variation to Pointwise Selection Principles*, SpringerBriefs in Optimization, https://doi.org/10.1007/978-3-030-87399-8_3

In fact, given $t \in T$, we have $i(\varphi) \le \varphi(t) \le s(\varphi)$, and so, subtracting $(i(\varphi)+s(\varphi))/2$ from both sides of these inequalities, we get

$$-\frac{|\varphi(T)|}{2} = \frac{i(\varphi) - s(\varphi)}{2} = i(\varphi) - \frac{i(\varphi) + s(\varphi)}{2} \le$$

$$\le \varphi(t) - \frac{i(\varphi) + s(\varphi)}{2} \le$$

$$\le s(\varphi) - \frac{i(\varphi) + s(\varphi)}{2} = \frac{s(\varphi) - i(\varphi)}{2} = \frac{|\varphi(T)|}{2}.$$

Since $|f(T)| = |\varphi(T)| \cdot \|x\|$, by (2.11), $V_\varepsilon(f, T) = 0$ for all $\varepsilon \ge |\varphi(T)| \cdot \|x\|$. Furthermore, if $c \equiv c(t) = (i(\varphi) + s(\varphi))(x/2) + y$, $t \in T$, then c is a constant function on T and, by (3.2), we have

$$\|f(t) - c\| = \left|\varphi(t) - \frac{i(\varphi) + s(\varphi)}{2}\right| \cdot \|x\| \le \frac{|\varphi(T)|}{2} \cdot \|x\| \quad \text{for all } t \in T,$$

i.e., $\|f - c\|_{\infty,T} \le |\varphi(T)| \cdot \|x\|/2$. By (2.10), we find

$$V_\varepsilon(f, T) = 0 \quad \text{for all} \quad \varepsilon \ge |\varphi(T)| \cdot \|x\|/2. \tag{3.3}$$

Now, assume that $0 < \varepsilon < |\varphi(T)| \cdot \|x\|/2$. Lemma 2.5(f) implies

$$V_\varepsilon(f, T) \ge |f(T)| - 2\varepsilon = |\varphi(T)| \cdot \|x\| - 2\varepsilon. \tag{3.4}$$

Define the function $g \in M^T$ by

$$g(t) = \left[\left(1 - \frac{2\varepsilon}{V(\varphi, T)\|x\|}\right)\varphi(t) + \frac{(i(\varphi) + s(\varphi))\varepsilon}{V(\varphi, T)\|x\|}\right]x + y = \tag{3.5}$$

$$= \varphi(t)x - \frac{2\varepsilon}{V(\varphi, T)\|x\|}\left(\varphi(t) - \frac{i(\varphi) + s(\varphi)}{2}\right)x + y, \quad t \in T. \tag{3.6}$$

Note that since $|\varphi(T)| \le V(\varphi, T)$, the assumption on ε gives $\varepsilon < V(\varphi, T)\|x\|/2$, and so, $0 < 2\varepsilon/(V(\varphi, T)\|x\|) < 1$. Given $t \in T$, (3.6) and (3.2) imply

$$\|f(t) - g(t)\| = \frac{2\varepsilon}{V(\varphi, T)\|x\|} \cdot \left|\varphi(t) - \frac{i(\varphi) + s(\varphi)}{2}\right| \cdot \|x\| \le \frac{2\varepsilon}{V(\varphi, T)} \cdot \frac{|\varphi(T)|}{2} \le \varepsilon,$$

and so, $\|f - g\|_{\infty,T} \le \varepsilon$. By (3.5), we find

$$V(g, T) = \left(1 - \frac{2\varepsilon}{V(\varphi, T)\|x\|}\right)V(\varphi, T)\|x\| = V(\varphi, T)\|x\| - 2\varepsilon.$$

Hence, by definition (2.6), $V_\varepsilon(f, T) \leq V(g, T) = V(\varphi, T)\|x\| - 2\varepsilon$. From here and (3.4) we conclude that

$$|\varphi(T)| \cdot \|x\| - 2\varepsilon \leq V_\varepsilon(f, T) \leq V(\varphi, T)\|x\| - 2\varepsilon \quad \text{if} \quad 0 < \varepsilon < |\varphi(T)| \cdot \|x\|/2.$$
(3.7)

In particular, if $\varphi \in \mathbb{R}^T$ is (nonconstant and) *monotone*, then $V(\varphi, T) = |\varphi(T)|$, and so, (3.7) yields:

$$\text{if} \quad 0 < \varepsilon < |\varphi(T)| \cdot \|x\|/2, \quad \text{then} \quad V_\varepsilon(f, T) = |\varphi(T)| \cdot \|x\| - 2\varepsilon.$$
(3.8)

Clearly, Example 2.7 is a particular case of (3.8) and (3.3) with $T = [0, 1]$, $M = \mathbb{R}$, $x = 1$, $y = 0$, and $\varphi(t) = t$, $t \in T$.

However, if φ from (3.1) is nonmonotone, both inequalities (3.7) may be strict (cf. Remark 3.4). Note also that assertion (3.8) implies the classical Helly selection principle for monotone functions (cf. Remark 4.5).

Example 3.2 Here we show that the inequalities in Lemma 2.8 are sharp and may be strict. In fact, letting $\varphi(t) = t$, $t \in T = [0, 1]$, and $y = 0$ in (3.1), and setting $T_1 = [0, \frac{1}{2}]$ and $T_2 = [\frac{1}{2}, 1]$, we get, by virtue of (3.8) and (3.3),

$$V_\varepsilon(f, T) = \begin{cases} \|x\| - 2\varepsilon & \text{if } 0 < \varepsilon < \frac{1}{2}\|x\|, \\ 0 & \text{if} \quad \varepsilon \geq \frac{1}{2}\|x\|, \end{cases}$$

and, for $i = 1, 2$,

$$V_\varepsilon(f, T_i) = \begin{cases} \frac{1}{2}\|x\| - 2\varepsilon & \text{if } 0 < \varepsilon < \frac{1}{4}\|x\|, \\ 0 & \text{if} \quad \varepsilon \geq \frac{1}{4}\|x\|. \end{cases}$$

It remains, in Lemma 2.8, to consider the following three cases: (a) $0 < \varepsilon < \frac{1}{4}\|x\|$, (b) $\frac{1}{4}\|x\| \leq \varepsilon < \frac{1}{2}\|x\|$, and (c) $\varepsilon \geq \frac{1}{2}\|x\|$. Explicitly, we have, in place of

$$V_\varepsilon(f, T_1) + V_\varepsilon(f, T_2) \leq V_\varepsilon(f, T) \leq V_\varepsilon(f, T_1) + V_\varepsilon(f, T_2) + 2\varepsilon :$$

(a) $\left(\frac{1}{2}\|x\| - 2\varepsilon\right) + \left(\frac{1}{2}\|x\| - 2\varepsilon\right) < \|x\| - 2\varepsilon = \left(\frac{1}{2}\|x\| - 2\varepsilon\right) + \left(\frac{1}{2}\|x\| - 2\varepsilon\right) + 2\varepsilon;$

(b) $0 + 0 < \|x\| - 2\varepsilon \leq 0 + 0 + 2\varepsilon;$

(c) $0 + 0 = \quad 0 \quad < 0 + 0 + 2\varepsilon.$

Example 3.3 Let $\tau \in I = [a, b]$, (M, d) be a metric space, and $x, y \in M, x \neq y$. Define $f \in M^I$ by

$$f(t) \equiv f_\tau(t) = \begin{cases} x & \text{if } t = \tau, \\ y & \text{if } t \in I, \ t \neq \tau. \end{cases} \tag{3.9}$$

Clearly, $|f(I)| = d(x, y)$, $V(f, I) = d(x, y)$ if $\tau \in \{a, b\}$, and $V(f, I) = 2d(x, y)$ if $a < \tau < b$. By (2.11), we get $V_\varepsilon(f, I) = 0$ for all $\varepsilon \geq d(x, y)$. Lemma 2.5(f) provides the following inequalities for $0 < \varepsilon < d(x, y)$:

(a) If $\tau = a$ or $\tau = b$, then

$$d(x, y) - 2\varepsilon \leq V_\varepsilon(f, I) \leq d(x, y) \quad \text{if} \quad 0 < \varepsilon < \tfrac{1}{2}d(x, y),$$

$$0 \leq V_\varepsilon(f, I) \leq d(x, y) \quad \text{if} \quad \tfrac{1}{2}d(x, y) \leq \varepsilon < d(x, y);$$

(b) If $a < \tau < b$, then

$$d(x, y) - 2\varepsilon \leq V_\varepsilon(f, I) \leq 2d(x, y) \quad \text{if} \quad 0 < \varepsilon < \tfrac{1}{2}d(x, y),$$

$$0 \leq V_\varepsilon(f, I) \leq 2d(x, y) \quad \text{if} \quad \tfrac{1}{2}d(x, y) \leq \varepsilon < d(x, y).$$

Under additional assumptions on the metric space (M, d), the values $V_\varepsilon(f, I)$ for $0 < \varepsilon < d(x, y)$ can be given more exactly. To see this, we consider two cases (A) and (B) below.

(A) Let $M = \{x, y\}$ be the two-point set with metric d and $0 < \varepsilon < d(x, y)$. Since $f(t) = x$ or $f(t) = y$ for all $t \in I$, we have: if $g \in M^I$ and $d_{\infty, I}(f, g) \leq \varepsilon$, then $g = f$ on I, i.e., $G_{\varepsilon, I}(f) = \{f\}$. Thus, $V_\varepsilon(f, I) = V(f, I)$, and so,

$$V_\varepsilon(f, I) = d(x, y) \quad \text{if} \quad \tau \in \{a, b\},$$

$$V_\varepsilon(f, I) = 2d(x, y) \quad \text{if} \quad a < \tau < b.$$

(B) Let $(M, \|\cdot\|)$ be a normed linear space with induced metric d and $0 < \varepsilon < d(x, y) = \|x - y\|$. By (2.12), $V_\varepsilon(f, I) = 0$ for all $\varepsilon \geq \tfrac{1}{2}\|x - y\|$. We assert that if $0 < \varepsilon < \tfrac{1}{2}\|x - y\|$, then

$$V_\varepsilon(f, I) = \|x - y\| - 2\varepsilon \quad \text{if} \quad \tau \in \{a, b\}, \tag{3.10}$$

$$V_\varepsilon(f, I) = 2(\|x - y\| - 2\varepsilon) \quad \text{if} \quad a < \tau < b. \tag{3.11}$$

In order to establish these equalities, we first note that the function f from (3.9) can be expressed as (cf. (3.1))

$$f(t) = \varphi(t)(x - y) + y, \quad \text{where } \varphi(t) \equiv \varphi_\tau(t) = \begin{cases} 1 & \text{if } t = \tau, \\ 0 & \text{if } t \neq \tau, \end{cases} \quad t \in I. \quad (3.12)$$

Proof of (3.10) If $\tau \in \{a, b\}$, then φ is monotone on I with $i(\varphi) = 0$, $s(\varphi) = 1$, and $V(\varphi, I) = |\varphi(I)| = 1$. Now, (3.10) follows from (3.12) and (3.8).

Note that function g from (3.5), used in obtaining (3.10), is of the form

$$g(t) = \left[\left(1 - \frac{2\varepsilon}{\|x - y\|}\right)\varphi(t) + \frac{\varepsilon}{\|x - y\|}\right](x - y) + y, \quad t \in I,$$

i.e., if $e_{x,y} = (x - y)/\|x - y\|$ is the *unit vector* ('directed from y to x'), then

$$g(\tau) = x - \varepsilon e_{x,y}, \quad \text{and} \quad g(t) = y + \varepsilon e_{x,y}, \quad t \in I \setminus \{\tau\}. \quad (3.13)$$

This implies $\|f - g\|_{\infty,I} = \varepsilon$ (for all $\tau \in I$), and we have, for $\tau \in \{a, b\}$,

$$V(g, I) = |g(I)| = \|(x - \varepsilon e_{x,y}) - (y + \varepsilon e_{x,y})\| = \|x - y\| - 2\varepsilon. \quad (3.14)$$

Proof of (3.11) Suppose $a < \tau < b$ and $0 < \varepsilon < \frac{1}{2}\|x - y\|$. First, consider an arbitrary function $g \in M^I$ such that $\|f - g\|_{\infty,I} = \sup_{t \in I} \|f(t) - g(t)\| \leq \varepsilon$. Since $P = \{a, \tau, b\}$ is a partition of I, by virtue of (2.2) and (3.9), we get

$$V(g, I) \geq \|g(\tau) - g(a)\| + \|g(b) - g(\tau)\|$$

$$\geq (\|f(\tau) - f(a)\| - 2\varepsilon) + (\|f(b) - f(\tau)\| - 2\varepsilon) \quad (3.15)$$

$$= 2(\|x - y\| - 2\varepsilon).$$

Due to the arbitrariness of g as above, (2.6) implies $V_\varepsilon(f, I) \geq 2(\|x - y\| - 2\varepsilon)$. Now, for the function g from (3.13), the additivity (**V.2**) of V and (3.14) yield

$$V(g, I) = V(g, [a, \tau]) + V(g, [\tau, b]) = |g([a, \tau])| + |g([\tau, b])| = 2(\|x - y\| - 2\varepsilon),$$

and so, $V_\varepsilon(f, I) \leq V(g, I) = 2(\|x - y\| - 2\varepsilon)$. This completes the proof of (3.11).

Remark 3.4 If φ from (3.1) is nonmonotone, inequalities in (3.7) may be *strict*. In fact, supposing $a < \tau < b$, we find that function $\varphi = \varphi_\tau$ from (3.12) is not monotone, $|\varphi(I)| = 1$, and $V(\varphi, I) = 2$, and so, by (3.11), inequalities (3.7) for function f from (3.12) are of the form:

$$\|x - y\| - 2\varepsilon < V_\varepsilon(f, I) = 2(\|x - y\| - 2\varepsilon) < 2\|x - y\| - 2\varepsilon$$

whenever $0 < \varepsilon < \frac{1}{2}\|x - y\|$.

Example 3.5 Let $I = [a, b]$, $a < \tau < b$, $(M, \|\cdot\|)$ be a normed linear space, $x, y \in M$, $x \neq y$, and $\alpha \in \mathbb{R}$. Define $f \in M^I$ by

$$f(t) = \begin{cases} x & \text{if } a \leq t < \tau, \\ (1-\alpha)x + \alpha y & \text{if } t = \tau, \\ y & \text{if } \tau < t \leq b. \end{cases} \tag{3.16}$$

We are going to evaluate the approximate variation $\{V_\varepsilon(f, I)\}_{\varepsilon > 0}$ for all $\alpha \in \mathbb{R}$. For this, we consider three possibilities: $0 \leq \alpha \leq 1$, $\alpha < 0$, and $\alpha > 1$.

Case $0 \leq \alpha \leq 1$ We assert that (independently of $\alpha \in [0, 1]$)

$$V_\varepsilon(f, I) = \begin{cases} \|x - y\| - 2\varepsilon & \text{if } 0 < \varepsilon < \frac{1}{2}\|x - y\|, \\ 0 & \text{if } \quad \varepsilon \geq \frac{1}{2}\|x - y\|. \end{cases} \tag{3.17}$$

To see this, we note that f can be represented in the form (3.1):

$$f(t) = \varphi(t)(x - y) + (1 - \alpha)x + \alpha y \quad \text{with} \quad \varphi(t) = \begin{cases} \alpha & \text{if } a \leq t < \tau, \\ 0 & \text{if } t = \tau, \\ \alpha - 1 & \text{if } \tau < t \leq b. \end{cases}$$

Since $\alpha \in [0, 1]$, φ is nonincreasing on I and $|\varphi(I)| = |\alpha - (\alpha - 1)| = 1$. Hence, (3.8) implies the first line in (3.17). The second line in (3.17) is a consequence of (3.3).

Case $\alpha < 0$ The resulting form of $V_\varepsilon(f, I)$ is given by (3.22), (3.24), and (3.20). Now we turn to their proofs. We set $x_\alpha = (1 - \alpha)x + \alpha y$ in (3.16) and note that

$$x_\alpha - x = (-\alpha)(x - y) = (-\alpha)\|x - y\|e_{x,y}, \tag{3.18}$$

$$x_\alpha - y = (1 - \alpha)(x - y) = (1 - \alpha)\|x - y\|e_{x,y}, \tag{3.19}$$

where $e_{x,y} = (x - y)/\|x - y\|$.

Let us evaluate $|f(I)|$ and $V(f, I)$. Since $1 - \alpha > -\alpha$, and $\alpha < 0$ implies $1 - \alpha > 1$, by (3.18) and (3.19), $\|x_\alpha - y\| > \|x_\alpha - x\|$ and $\|x_\alpha - y\| > \|x - y\|$, and since f assumes only values x, x_α, and y,

$$|f(I)| = \|x_\alpha - y\| = (1 - \alpha)\|x - y\|.$$

For $V(f, I)$, by the additivity (**V.2**) of V, (3.18), and (3.19), we find

$$V(f, I) = V(f, [a, \tau]) + V(f, [\tau, b]) = |f([a, \tau])| + |f([\tau, b])|$$

$$= \|f(\tau) - f(a)\| + \|f(b) - f(\tau)\| = \|x_\alpha - x\| + \|y - x_\alpha\|$$

$$= (-\alpha)\|x - y\| + (1 - \alpha)\|x - y\| = (1 - 2\alpha)\|x - y\|.$$

Setting $c = c(t) = \frac{1}{2}(x_\alpha + y)$ for all $t \in I$, we get, by (3.19),

$$\|x_\alpha - c\| = \|y - c\| = \frac{1}{2}\|x_\alpha - y\| = \frac{1}{2}(1-\alpha)\|x - y\| = \frac{1}{2}|f(I)|,$$

and

$$\|x - c\| = \|x - \frac{1}{2}(x_\alpha + y)\| = \frac{1}{2}\|(x - x_\alpha) + (x - y)\| = \frac{1}{2}\|\alpha(x - y) + (x - y)\|$$

$$= \frac{1}{2}|\alpha + 1| \cdot \|x - y\| \leq \frac{1}{2}(1 + |\alpha|)\|x - y\| \overset{(\alpha < 0)}{=} \frac{1}{2}(1 - \alpha)\|x - y\| = \frac{1}{2}|f(I)|.$$

Hence $\|f - c\|_{\infty, I} \leq \frac{1}{2}|f(I)|$, and it follows from (2.10) that

$$V_\varepsilon(f, I) = 0 \quad \text{if} \quad \varepsilon \geq \frac{1}{2}|f(I)| = \frac{1-\alpha}{2}\|x - y\|. \tag{3.20}$$

It remains to consider the case when $0 < \varepsilon < \frac{1-\alpha}{2}\|x - y\|$, which we split into two subcases:

(I) $0 < \varepsilon < \frac{(-\alpha)}{2}\|x - y\|$, and (II) $\frac{(-\alpha)}{2}\|x - y\| \leq \varepsilon < \frac{1-\alpha}{2}\|x - y\|$.

Subcase (I) First, given $g \in M^I$ with $\|f - g\|_{\infty, I} \leq \varepsilon$, since $P = \{a, \tau, b\}$ is a partition of I, applying (3.15), we get

$$V(g, I) \geq (\|f(\tau) - f(a)\| - 2\varepsilon) + (\|f(b) - f(\tau)\| - 2\varepsilon)$$

$$= ((-\alpha)\|x - y\| - 2\varepsilon) + ((1 - \alpha)\|x - y\| - 2\varepsilon)$$

$$= (1 - 2\alpha)\|x - y\| - 4\varepsilon,$$

and so, by (2.6), $V_\varepsilon(f, I) \geq (1 - 2\alpha)\|x - y\| - 4\varepsilon$. Now, we define a concrete (="test") function $g \in M^I$ by the rule:

$$g(t) = x + \varepsilon e_{x,y} \quad \text{if} \quad a \leq t < \tau,$$

$$g(\tau) = x_\alpha - \varepsilon e_{x,y}, \tag{3.21}$$

$$g(t) = y + \varepsilon e_{x,y} \quad \text{if} \quad \tau < t \leq b.$$

Clearly, by (3.16) and (3.21), $\|f - g\|_{\infty, I} = \varepsilon$. Furthermore,

$$V(g, I) = \|g(\tau) - g(a)\| + \|g(\tau) - g(b)\|$$

$$= \|(x_\alpha - x) - 2\varepsilon e_{x,y}\| + \|(x_\alpha - y) - 2\varepsilon e_{x,y}\|$$

$$= \left\|(-\alpha)\|x - y\|e_{x,y} - 2\varepsilon e_{x,y}\right\| + \left\|(1 - \alpha)\|x - y\|e_{x,y} - 2\varepsilon e_{x,y}\right\|$$

$$= \left|(-\alpha)\|x - y\| - 2\varepsilon\right| + \left|(1 - \alpha)\|x - y\| - 2\varepsilon\right|.$$

Assumption (I) implies $2\varepsilon < (-\alpha)\|x - y\| < (1 - \alpha)\|x - y\|$, so

$$V(g, I) = ((-\alpha)\|x - y\| - 2\varepsilon) + ((1 - \alpha)\|x - y\| - 2\varepsilon) = (1 - 2\alpha)\|x - y\| - 4\varepsilon.$$

By (2.6), $V_\varepsilon(f, I) \leq V(g, I) = (1 - 2\alpha)\|x - y\| - 4\varepsilon$. Thus,

$$V_\varepsilon(f, I) = (1 - 2\alpha)\|x - y\| - 4\varepsilon \quad \text{if} \quad 0 < \varepsilon < \tfrac{(-\alpha)}{2}\|x - y\|. \tag{3.22}$$

Note that, in agreement with Lemma 2.5(a), $V_\varepsilon(f, I) \to V(f, I)$ as $\varepsilon \to +0$.

Subcase (II) First, given $g \in M^I$ with $\|f - g\|_{\infty,I} \leq \varepsilon$, by virtue of (2.2) and (3.19), we get

$$V(g, I) \geq \|g(b) - g(\tau)\| \geq \|f(b) - f(\tau)\| - 2\varepsilon = (1 - \alpha)\|x - y\| - 2\varepsilon,$$

and so, definition (2.6) implies $V_\varepsilon(f, I) \geq (1 - \alpha)\|x - y\| - 2\varepsilon$. Now, define a test function $g \in M^I$ by

$$g(t) = \begin{cases} x_\alpha - \varepsilon e_{x,y} & \text{if } a \leq t \leq \tau, \\ y + \varepsilon e_{x,y} & \text{if } \tau < t \leq b. \end{cases} \tag{3.23}$$

Let us show that $\|f - g\|_{\infty,I} \leq \varepsilon$. Clearly, by (3.16), $\|f(t) - g(t)\| = \varepsilon$ for all $\tau \leq t \leq b$. Now, suppose $a \leq t < \tau$. We have, by (3.18),

$$\|f(t) - g(t)\| = \|x - x_\alpha + \varepsilon e_{x,y}\| = \left\| \alpha\|x - y\| e_{x,y} + \varepsilon e_{x,y} \right\|$$

$$= \left| \alpha\|x - y\| + \varepsilon \right| \equiv A_\alpha.$$

Suppose first that $\alpha > -1$ (that is, x_α is closer to x than x to y in the sense that $\|x_\alpha - x\| = (-\alpha)\|x - y\| < \|x - y\|$). Then $(-\alpha) < \frac{1}{2}(1 - \alpha)$, and so, for ε from subcase (II) we have either

(II$_1$) $\tfrac{(-\alpha)}{2}\|x - y\| \leq \varepsilon < (-\alpha)\|x - y\|$, or (II$_2$) $(-\alpha)\|x - y\| \leq \varepsilon < \tfrac{1-\alpha}{2}\|x - y\|$.

In case (II$_1$), $\alpha\|x - y\| + \varepsilon < 0$, which implies $A_\alpha = (-\alpha)\|x - y\| - \varepsilon$. Hence, the left-hand side inequality in (II$_1$) gives $A_\alpha \leq \varepsilon$. In case (II$_2$), $\alpha\|x - y\| + \varepsilon \geq 0$, which implies $A_\alpha = \alpha\|x - y\| + \varepsilon < \varepsilon$ (because $\alpha < 0$).

Now, suppose $\alpha \leq -1$ (i.e., $\|x - y\| \leq (-\alpha)\|x - y\| = \|x_\alpha - x\|$, which means that x_α is farther from x than x from y), so that $\frac{1}{2}(1 - \alpha) \leq (-\alpha)$. In this case, assumption (II) implies only condition (II$_1$), and so, as above, $A_\alpha = (-\alpha)\|x - y\| - \varepsilon \leq \varepsilon$. This completes the proof of the inequality $\|f - g\|_{\infty,I} \leq \varepsilon$.

For the variation $V(g, I)$ of function g from (3.23), we have, by (3.19),

$$V(g, I) = \|(x_\alpha - \varepsilon e_{x,y}) - (y + \varepsilon e_{x,y})\| = \|(x_\alpha - y) - 2\varepsilon e_{x,y}\|$$

$$= \left\|(1 - \alpha)\|x - y\|e_{x,y} - 2\varepsilon e_{x,y}\right\| = (1 - \alpha)\|x - y\| - 2\varepsilon.$$

Hence $V_\varepsilon(f, I) \leq V(g, I) = (1 - \alpha)\|x - y\| - 2\varepsilon$. Thus, we have shown that

$$V_\varepsilon(f, I) = (1 - \alpha)\|x - y\| - 2\varepsilon \quad \text{if} \quad \tfrac{(-\alpha)}{2}\|x - y\| \leq \varepsilon < \tfrac{(1-\alpha)}{2}\|x - y\|. \quad (3.24)$$

Case $\alpha > 1$ We reduce this case to the case $\alpha < 0$ and apply Lemma 2.13. Set $T = [a', b']$ with $a' = 2\tau - b$ and $b' = 2\tau - a$, so that $a' < \tau < b'$, and define $\varphi : T \to \mathbb{R}$ by $\varphi(t) = 2\tau - t$, $a' \leq t \leq b'$. Clearly, φ is strictly decreasing on T, $\varphi(T) = [a, b] = I$, and $\varphi(\tau) = \tau$. Let us show that the composed function $f' = f \circ \varphi \in M^T$ is of the same form as (3.16).

If $a' \leq t < \tau$, then $\tau < \varphi(t) \leq b$, and so, by (3.16), $f'(t) = f(\varphi(t)) = y$; if $t = \tau$, then $f'(\tau) = f(\varphi(\tau)) = f(\tau) = x_\alpha$; and if $\tau < t \leq b'$, then $a \leq \varphi(t) < \tau$, and so, $f'(t) = f(\varphi(t)) = x$. Setting $x' = y$, $y' = x$, and $\alpha' = 1 - \alpha$, we get $\alpha' < 0$,

$$f'(t) = x' \text{ if } a' \leq t < \tau, \quad f'(t) = y' \text{ if } \tau < t \leq b',$$

and

$$f'(\tau) = x_\alpha = (1 - \alpha)x + \alpha y = \alpha'y' + (1 - \alpha')x' = (1 - \alpha')x' + \alpha'y' \equiv x'_{\alpha'}.$$

By Lemma 2.13, given $\varepsilon > 0$,

$$V_\varepsilon(f, I) = V_\varepsilon(f, [a, b]) = V_\varepsilon(f, \varphi(T)) = V_\varepsilon(f \circ \varphi, T) = V_\varepsilon(f', [a', b']),$$

where, since f' is of the form (3.16), $V_\varepsilon(f', [a', b'])$ is given by (3.22), (3.24), and (3.20) with f, x, y, and α replaced by f', x', y', and α', respectively. Noting that $\|x' - y'\| = \|x - y\|$, $1 - \alpha' = \alpha$, $1 - 2\alpha' = 2\alpha - 1$, and $(-\alpha') = \alpha - 1$, we get, in the case when $\alpha > 1$:

$$V_\varepsilon(f, I) = \begin{cases} (2\alpha - 1)\|x - y\| - 4\varepsilon & \text{if} & 0 < \varepsilon < \tfrac{\alpha-1}{2}\|x - y\|, \\ \alpha\|x - y\| - 2\varepsilon & \text{if } \tfrac{\alpha-1}{2}\|x - y\| \leq \varepsilon < \tfrac{\alpha}{2}\|x - y\|, \\ 0 & \text{if} & \varepsilon \geq \tfrac{\alpha}{2}\|x - y\|. \end{cases}$$

Finally, we note that, for $\alpha > 1$, we have, by (3.18) and (3.19),

$$V(f, I) = \|x - x_\alpha\| + \|x_\alpha - y\| = \alpha\|x - y\| + (\alpha - 1)\|x - y\| = (2\alpha - 1)\|x - y\|,$$

and so, $V_\varepsilon(f, I) \to V(f, I)$ as $\varepsilon \to +0$.

3.2 Generalized Dirichlet's function

Example 3.6 (Generalized Dirichlet's function) This is an illustration of Lemma 2.10 illuminating several specific features of the approximate variation.

(a) Let $T = I = [a, b]$, (M, d) be a metric space, and $\mathbb{Q}$ denote (as usual) the set of all rational numbers. We set $I_1 = I \cap \mathbb{Q}$ and $I_2 = I \setminus \mathbb{Q}$. A function $f \in M^I$ is said to be a *generalized Dirichlet function* if $f \in B(I; M)$ and

$$\Delta f \equiv \Delta f(I_1, I_2) = \inf_{s \in I_1, t \in I_2} d(f(s), f(t)) > 0.$$

Clearly, $f \notin \mathrm{Reg}(I; M)$ (in fact, if, say, $a < \tau \le b$, then for all $\delta \in (0, \tau - a)$, $s \in (\tau - \delta, \tau) \cap \mathbb{Q}$, and $t \in (\tau - \delta, \tau) \setminus \mathbb{Q}$, we have $d(f(s), f(t)) \ge \Delta f > 0$).

Setting $|f(I_1, I_2)| = \sup_{s \in I_1, t \in I_2} d(f(s), f(t))$, we find

$$|f(I_1, I_2)| \le |f(I_1)| + d(f(s_0), f(t_0)) + |f(I_2)|, \quad s_0 \in I_1, \ t_0 \in I_2,$$

and

$$0 < \Delta f \le |f(I_1, I_2)| \le |f(I)| = \max\{|f(I_1)|, |f(I_2)|, |f(I_1, I_2)|\}.$$

Furthermore (cf. Lemma 2.10), we have

$$V_\varepsilon(f, I) = \infty \ \text{ if } \ 0 < \varepsilon < \Delta f/2, \ \text{ and } \ V_\varepsilon(f, I) = 0 \ \text{ if } \ \varepsilon \ge |f(I)|; \tag{3.25}$$

the values of $V_\varepsilon(f, I)$ for $\Delta f/2 \le \varepsilon < |f(I)|$ depend on (the structure of) the metric space (M, d) in general (see items (b), (c), and (d) below). The second equality in (3.25) is a consequence of (2.11). In order to prove the first assertion in (3.25), we show that if $0 < \varepsilon < \Delta f/2$, $g \in M^I$, and $d_{\infty, I}(f, g) \le \varepsilon$, then $V(g, I) = \infty$ (cf. (2.7)). In fact, given $n \in \mathbb{N}$, let $P = \{t_i\}_{i=0}^{2n}$ be a partition of I (i.e., $a \le t_0 < t_1 < t_2 < \cdots < t_{2n-1} < t_{2n} \le b$) such that $\{t_{2i}\}_{i=0}^{n} \subset I_1$ and $\{t_{2i-1}\}_{i=1}^{n} \subset I_2$. Given $i \in \{1, 2, \ldots, n\}$, by the triangle inequality for d, we have

$$d(f(t_{2i}), f(t_{2i-1})) \le d(f(t_{2i}), g(t_{2i})) + d(g(t_{2i}), g(t_{2i-1})) + d(g(t_{2i-1}), f(t_{2i-1}))$$

$$\le d_{\infty, I_1}(f, g) + d(g(t_{2i}), g(t_{2i-1})) + d_{\infty, I_2}(g, f)$$

$$\le \varepsilon + d(g(t_{2i}), g(t_{2i-1})) + \varepsilon. \tag{3.26}$$

It follows from the definition of Jordan's variation $V(g, I)$ that

$$V(g, I) \geq \sum_{i=1}^{2n} d(g(t_i), g(t_{i-1})) \geq \sum_{i=1}^{n} d(g(t_{2i}), g(t_{2i-1}))$$

$$\geq \sum_{i=1}^{n} \left(d(f(t_{2i}), f(t_{2i-1})) - 2\varepsilon \right) \geq (\Delta f - 2\varepsilon)n. \qquad (3.27)$$

It remains to take into account the arbitrariness of $n \in \mathbb{N}$.

In a particular case of the classical *Dirichlet function* $f = \mathscr{D}_{x,y} : I \to M$ defined, for $x, y \in M$, $x \neq y$, by

$$\mathscr{D}_{x,y}(t) = \begin{cases} x & \text{if } t \in I_1, \\ y & \text{if } t \in I_2, \end{cases} \qquad (3.28)$$

we have $\Delta f = \Delta \mathscr{D}_{x,y} = d(x, y)$ and $|f(I)| = |\mathscr{D}_{x,y}(I_1, I_2)| = d(x, y)$, and so, (3.25) assumes the form (which was established in [29, assertion (4.4)]):

$$V_\varepsilon(f, I) = \infty \text{ if } 0 < \varepsilon < d(x, y)/2, \text{ and } V_\varepsilon(f, I) = 0 \text{ if } \varepsilon \geq d(x, y). \qquad (3.29)$$

(b) This example and items (c) and (d) below illustrate the sharpness of assertions in Lemma 2.5(b), (d). Let $(M, \| \cdot \|)$ be a normed linear space with induced metric d (cf. p. 11) and $f = \mathscr{D}_{x,y}$ be the Dirichlet function (3.28). Setting $c = c(t) = (x + y)/2$, $t \in I$, we find

$$2d_{\infty,I}(f, c) = 2 \max\{\|x - c\|, \|y - c\|\} = \|x - y\| = d(x, y) = |f(I)|,$$

and so, by (2.12) and (2.13), the second equality in (3.29) is refined as follows:

$$V_\varepsilon(f, I) = 0 \quad \text{for all} \quad \varepsilon \geq \frac{\|x - y\|}{2} = \frac{d(x, y)}{2}. \qquad (3.30)$$

This shows the sharpness of the inequality in Lemma 2.5(b). In Lemma 2.5(d), inequalities assume the form:

$$\inf_{\varepsilon > 0}(V_\varepsilon(f, I) + \varepsilon) = \frac{\|x - y\|}{2} < |f(I)| = \|x - y\| = \inf_{\varepsilon > 0}(V_\varepsilon(f, I) + 2\varepsilon).$$

More generally, (3.29) and (3.30) hold for a complete and *metrically convex* (in the sense of K. Menger [57]) metric space (M, d) (see [29, Example 1]).

(c) In the context of (3.28), assume that $M = \{x, y\}$ is the two-point set with metric d. If $0 < \varepsilon < d(x, y)$, $g \in M^I$, and $d_{\infty,I}(f, g) \leq \varepsilon$, then $g = f = \mathscr{D}_{x,y}$ on I, and so, $V(g, I) = \infty$. By (2.7), the first assertion in (3.29) can be expressed more exactly as $V_\varepsilon(f, I) = \infty$ for all $0 < \varepsilon < d(x, y)$. Now, (in)equalities in Lemma 2.5(d) are of the form:

$$\inf_{\varepsilon > 0}(V_\varepsilon(f, I) + \varepsilon) = d(x, y) = |f(I)| < 2d(x, y) = \inf_{\varepsilon > 0}(V_\varepsilon(f, I) + 2\varepsilon).$$

(d) Given $x, y \in \mathbb{R}$, $x \neq y$, and $0 \leq r \leq |x - y|/2$, we set

$$M_r = \mathbb{R} \setminus \left(\tfrac{1}{2}(x + y) - r, \tfrac{1}{2}(x + y) + r\right) \quad \text{and} \quad d(u, v) = |u - v|, \ u, v \in M_r.$$

Note that (M_r, d) is a proper metric space (cf. p. 21). If $f = \mathscr{D}_{x,y} : I \to M_r$ is the Dirichlet function (3.28) on I, we claim that

$$V_\varepsilon(f, I) = \infty \text{ if } 0 < \varepsilon < \frac{1}{2}|x - y| + r, \text{ and } V_\varepsilon(f, I) = 0 \text{ otherwise.}$$
$$\tag{3.31}$$

Proof of (3.31) Since $M_0 = \mathbb{R}$, assertion (3.31) for $r = 0$ follows from (3.29) and (3.30). Now, suppose $r > 0$. From (3.29), we find $V_\varepsilon(f, I) = \infty$ if $0 < \varepsilon < \tfrac{1}{2}|x - y|$, and $V_\varepsilon(f, I) = 0$ if $\varepsilon \geq |x - y|$. So, only the case when $\tfrac{1}{2}|x - y| \leq \varepsilon < |x - y|$ is to be considered. We split this case into two subcases:

(I) $\tfrac{1}{2}|x - y| \leq \varepsilon < \tfrac{1}{2}|x - y| + r$, and (II) $\tfrac{1}{2}|x - y| + r \leq \varepsilon < |x - y|$.

Case (I) Let us show that if $g : I \to M_r$ and $d_{\infty, I}(f, g) \leq \varepsilon$, then $V(g, I) = \infty$. Given $t \in I = I_1 \cup I_2$, the inclusion $g(t) \in M_r$ is equivalent to

$$g(t) \leq \tfrac{1}{2}(x + y) - r \quad \text{or} \quad g(t) \geq \tfrac{1}{2}(x + y) + r, \tag{3.32}$$

and condition $d_{\infty, I}(f, g) = |f - g|_{\infty, I} \leq \varepsilon$ is equivalent to

$$|x - g(s)| \leq \varepsilon \ \forall s \in I_1, \text{ and } |y - g(t)| \leq \varepsilon \ \forall t \in I_2. \tag{3.33}$$

Due to the symmetry in x and y everywhere, we may assume that $x < y$.

Suppose $s \in I_1$. The first condition in (3.33) and assumption (I) imply

$$x - \varepsilon \leq g(s) \leq x + \varepsilon < x + \tfrac{1}{2}|x - y| + r = x + \tfrac{1}{2}(y - x) + r = \tfrac{1}{2}(x + y) + r,$$

and so, by (3.32), we find $g(s) \leq \tfrac{1}{2}(x + y) - r$. Note that, by (I),

$$-\varepsilon \leq g(s) - x \leq \tfrac{1}{2}(x + y) - r - x = \tfrac{1}{2}(y - x) - r = \tfrac{1}{2}|y - x| - r \leq \varepsilon - r < \varepsilon.$$

Given $t \in I_2$, the second condition in (3.33) and assumption (I) yield

$$y + \varepsilon \geq g(t) \geq y - \varepsilon > y - \tfrac{1}{2}|x - y| - r = y - \tfrac{1}{2}(y - x) - r = \tfrac{1}{2}(x + y) - r,$$

and so, by (3.32), we get $g(t) \geq \frac{1}{2}(x + y) + r$. Note also that, by (I),

$$\varepsilon \geq g(t) - y \geq \tfrac{1}{2}(x+y) + r - y = \tfrac{1}{2}(x-y) + r = -\tfrac{1}{2}|x-y| + r \geq -\varepsilon + r > -\varepsilon.$$

Thus, we have shown that, given $s \in I_1$ and $t \in I_2$,

$$g(t) - g(s) \geq \tfrac{1}{2}(x + y) + r - \left(\tfrac{1}{2}(x + y) - r\right) = 2r. \tag{3.34}$$

Given $n \in \mathbb{N}$, let $\{t_i\}_{i=0}^{2n}$ be a partition of I such that $\{t_{2i}\}_{i=0}^{n} \subset I_1$ and $\{t_{2i-1}\}_{i=1}^{n} \subset I_2$. Taking into account (3.34) with $s = t_{2i}$ and $t = t_{2i-1}$, we get

$$V(g, I) \geq \sum_{i=1}^{2n} |g(t_i) - g(t_{i-1})| \geq \sum_{i=1}^{n} \left(g(t_{2i-1}) - g(t_{2i})\right) \geq 2rn.$$

Case (II) We set $c = c(t) = \varepsilon + \min\{x, y\}$, $t \in I$; under our assumption $x < y$, we have $c = \varepsilon + x$. Note that $c \in M_r$: in fact, (II) and $x < y$ imply $\frac{1}{2}(y - x) + r \leq \varepsilon < y - x$, and so, $\frac{1}{2}(x + y) + r \leq c = \varepsilon + x < y$. If $s \in I_1$, we find $|x - c(s)| = \varepsilon$, and if $t \in I_2$, we get, by assumption (II),

$$|y - c(t)| = |y - x - \varepsilon| = y - x - \varepsilon \leq |x - y| - \tfrac{1}{2}|x - y| - r \leq \tfrac{1}{2}|x - y| + r \leq \varepsilon.$$

It follows that (cf. (3.33)) $d_{\infty, I}(f, c) \leq \varepsilon$, and since c is constant on I, we conclude from (2.10) that $V_\varepsilon(f, I) = 0$. This completes the proof of (3.31).
□

Two conclusions from (3.31) are in order. First, given $0 \leq r \leq \frac{1}{2}|x - y|$ and $\varepsilon > 0$, $V_\varepsilon(f, I) = 0$ if and only if $|f(I)| = |x - y| \leq 2\varepsilon - 2r$ (cf. (2.11) and Lemma 2.5(b)). Second, the inequalities in Lemma 2.5(d) are as follows:

$$\inf_{\varepsilon>0}(V_\varepsilon(f, I) + \varepsilon) = \tfrac{1}{2}|x - y| + r \leq |f(I)| = |x - y|$$

$$\leq |x - y| + 2r = \inf_{\varepsilon>0}(V_\varepsilon(f, I) + 2\varepsilon).$$

The inequalities at the left and at the right become equalities for $r = \frac{1}{2}|x - y|$ and $r = 0$, respectively; otherwise, the mentioned inequalities are strict.

(e) Let $x, y \in \mathbb{R}$, $x \neq y$, and $0 \leq r < |x - y|/2$. We set

$$M_r = \mathbb{R} \setminus \left[\tfrac{1}{2}(x + y) - r, \tfrac{1}{2}(x + y) + r\right] \quad \text{and} \quad d(u, v) = |u - v|, \; u, v \in M_r.$$

Note that (M_r, d) is an improper metric space. For the Dirichlet function $f = \mathscr{D}_{x,y} : I \to M_r$ from (3.28), we have:

$$V_\varepsilon(f, I) = \infty \text{ if } 0 < \varepsilon \le \frac{1}{2}|x - y| + r, \text{ and } V_\varepsilon(f, I) = 0 \text{ otherwise.}$$
(3.35)

Clearly, the function $\varepsilon \mapsto V_\varepsilon(f, I)$ is not continuous from the right at $\varepsilon = \frac{1}{2}|x - y| + r$ (cf. Lemma 2.14(a)). The proof of (3.35) follows the same lines as those of (3.31), so we present only the necessary modifications. We split the case when $\frac{1}{2}|x - y| \le \varepsilon < |x - y|$ into two subcases:

(I) $\frac{1}{2}|x - y| \le \varepsilon \le \frac{1}{2}|x - y| + r$, and (II) $\frac{1}{2}|x - y| + r < \varepsilon < |x - y|$.

Case (I) Given $g : I \to M_r$ with $d_{\infty,I}(f, g) \le \varepsilon$, to see that $V(g, I) = \infty$, we have *strict* inequalities in (3.32), conditions (3.33), and assume that $x < y$. If $s \in I_1$, then (as above) $g(s) \le \frac{1}{2}(x + y) + r$, and so, by (strict) (3.32), $g(s) < \frac{1}{2}(x + y) - r$. If $t \in I_2$, then $g(t) \ge \frac{1}{2}(x + y) - r$, and so, by (3.32), $g(t) > \frac{1}{2}(x+y)+r$. Thus, g is discontinuous at every point of $I = I_1 \cup I_2$, and so, $V(g, I) = \infty$. In fact, if, on the contrary, g is continuous at, say, a point $s \in I_1$, then the inequality $g(s) < \frac{1}{2}(x + y) - r$ holds in a neighborhood of s, and since the neighborhood contains an irrational point $t \in I_2$, we get $g(t) > \frac{1}{2}(x+y)+r$, which is a contradiction; recall also that a $g \in \mathrm{BV}(I; \mathbb{R})$ is continuous on I apart, possibly, an at most countable subset of I.

Case (II) It is to be noted only that $\frac{1}{2}(x + y) + r < c = \varepsilon + x < y$, and so, $c \in M_r$; in fact, by (II) and assumption $x < y$, $\frac{1}{2}(y - x) + r < \varepsilon < y - x$. □

3.3 Examples with convergent sequences

Example 3.7 The left limit $V_{\varepsilon-0}(f, T)$ in Lemma 2.12(a) cannot, in general, be replaced by $V_\varepsilon(f, T)$. To see this, we let $T = I$, $(M, \|\cdot\|)$ be a normed linear space, $\{x_j\}, \{y_j\} \subset M$ be two sequences, $x, y \in M$, $x \ne y$, and $x_j \to x$ and $y_j \to y$ in M as $j \to \infty$. If $f_j = \mathscr{D}_{x_j,y_j}$, $j \in \mathbb{N}$, and $f = \mathscr{D}_{x,y}$ are Dirichlet's functions (3.28) on I, then $f_j \rightrightarrows f$ on I, which follows from

$$\|f_j - f\|_{\infty,I} = \max\{\|x_j - x\|, \|y_j - y\|\} \to 0 \quad \text{as} \quad j \to \infty.$$

The values $V_\varepsilon(f, I)$ are given by (3.29) and (3.30), and, similarly, if $j \in \mathbb{N}$,

$$V_\varepsilon(f_j, I) = \infty \text{ if } 0 < \varepsilon < \frac{1}{2}\|x_j - y_j\|, \quad V_\varepsilon(f_j, I) = 0 \text{ if } \varepsilon \ge \frac{1}{2}\|x_j - y_j\|.$$
(3.36)

Setting $\varepsilon = \frac{1}{2}\|x - y\|$, $\alpha_j = 1 + (1/j)$, $x_j = \alpha_j x$, and $y_j = \alpha_j y$, $j \in \mathbb{N}$, we find

$$V_{\varepsilon+0}(f, I) = V_\varepsilon(f, I) = 0 < \infty = V_{\varepsilon-0}(f, I),$$

whereas, since $\varepsilon < \frac{1}{2}\alpha_j\|x - y\| = \frac{1}{2}\|x_j - y_j\|$ for all $j \in \mathbb{N}$,

$$V_\varepsilon(f_j, I) = \infty \text{ for all } j \in \mathbb{N}, \text{ and so, } \lim_{j\to\infty} V_\varepsilon(f_j, I) = \infty.$$

Example 3.8 The right-hand side inequality in Lemma 2.12(a) may not hold if $\{f_j\} \subset M^T$ converges to $f \in M^T$ only *pointwise* on T. To see this, suppose $C \equiv \inf_{j\in\mathbb{N}} |f_j(T)| > 0$ and $f = c$ (is a constant function) on T. Given $0 < \varepsilon < C/2$, Lemma 2.5(f) implies

$$V_\varepsilon(f_j, T) \geq |f_j(T)| - 2\varepsilon \geq C - 2\varepsilon > 0 = V_\varepsilon(c, T) = V_{\varepsilon-0}(f, T), \quad j \in \mathbb{N}.$$

For instance, given a sequence $\{\tau_j\} \subset (a, b) \subset I = [a, b]$ such that $\tau_j \to a$ as $j \to \infty$, and $x, y \in M$, $x \neq y$, defining $\{f_j\} \subset M^I$ (as in Example 3.3) by $f_j(\tau_j) = x$ and $f_j(t) = y$ if $t \in I \setminus \{\tau_j\}$, $j \in \mathbb{N}$, we have $C = d(x, y) > 0$ and $f_j \to c \equiv y$ pointwise on I.

The arguments above are not valid for the uniform convergence: in fact, if $f_j \rightrightarrows f = c$ on T, then, by (2.3), $|f_j(T)| \leq 2d_{\infty,T}(f_j, c) \to 0$ as $j \to \infty$, and so, $C = 0$.

Example 3.9 Lemma 2.12(b) is wrong for the pointwise convergence $f_j \to f$. To see this, let $T = I = [a, b]$, (M, d) be a metric space, $x, y \in M$, $x \neq y$, and, given $j \in \mathbb{N}$, define $f_j \in M^I$ at $t \in I$ by: $f_j(t) = x$ if $j!t$ is an integer, and $f_j(t) = y$ otherwise. Each f_j is a step function on I, so it is regulated and, hence, by Lemma 2.10, $V_\varepsilon(f_j, I) < \infty$ for all $\varepsilon > 0$. At the same time, the sequence $\{f_j\}$ converges (only) pointwise on I to the Dirichlet function $f = \mathscr{D}_{x,y}$ (cf. (3.28)), and so, by (3.29), $V_\varepsilon(f, I) = \infty$ for all $0 < \varepsilon < \frac{1}{2}d(x, y)$.

3.4 Examples with improper metric spaces

Example 3.10 This example is similar to Example 3.6(e) (p. 37), but with *finite* values of $V_\varepsilon(f, I)$. It shows that the assumption on the *proper* metric space (M, d) in Lemma 2.14(b) is essential.

Let $x, y \in \mathbb{R}$, $x \neq y$, $M = \mathbb{R} \setminus \{\frac{1}{2}(x + y)\}$ with metric $d(u, v) = |u - v|$ for $u, v \in M$, $I = [a, b]$, $\tau = a$ or $\tau = b$, and $f \in M^I$ be given by (cf. (3.9)): $f(\tau) = x$ and $f(t) = y$ if $t \in I$, $t \neq \tau$. We claim that (as in (3.10))

$$V_\varepsilon(f, I) = \begin{cases} |x - y| - 2\varepsilon & \text{if } 0 < \varepsilon < \frac{1}{2}|x - y|, \\ 0 & \text{if } \varepsilon \geq \frac{1}{2}|x - y|. \end{cases} \tag{3.37}$$

In order to verify this, we note that $|f(I)| = |x - y|$, and so, by (2.11), $V_\varepsilon(f, I) = 0$ for all $\varepsilon \geq |x - y|$. We split the case $0 < \varepsilon < |x - y|$ into the three possibilities:

$$\text{(I) } 0 < \varepsilon < \tfrac{1}{2}|x - y|; \text{ (II) } \varepsilon = \tfrac{1}{2}|x - y|; \text{ (III) } \tfrac{1}{2}|x - y| < \varepsilon < |x - y|.$$

Due to the symmetry (in x and y), we may consider only the case when $x < y$.

Case (I) Given $g \in M^I$ with $d_{\infty,I}(f, g) \leq \varepsilon$, inequality (2.2) implies

$$V(g, I) \geq |g(t) - g(\tau)| \geq |f(t) - f(\tau)| - 2\varepsilon = |x - y| - 2\varepsilon \quad (t \neq \tau),$$

and so, by (2.6), $V_\varepsilon(f, I) \geq |x - y| - 2\varepsilon$. Now, following (3.13), we set

$$g_\varepsilon(\tau) = x + \varepsilon \quad \text{and} \quad g_\varepsilon(t) = y - \varepsilon \ \text{ if } \ t \in I \setminus \{\tau\}. \tag{3.38}$$

We have $g_\varepsilon : I \to M$, because assumption $0 < \varepsilon < \tfrac{1}{2}(y - x)$ yields

$$g_\varepsilon(\tau) = x + \varepsilon < x + \tfrac{1}{2}(y - x) = \tfrac{1}{2}(x + y)$$

and, if $t \in I, t \neq \tau$,

$$g_\varepsilon(t) = y - \varepsilon > y - \tfrac{1}{2}(y - x) = \tfrac{1}{2}(x + y).$$

Moreover, $d_{\infty,I}(f, g_\varepsilon) = \varepsilon$ and

$$V(g_\varepsilon, I) = |g_\varepsilon(I)| = |(y - \varepsilon) - (x + \varepsilon)| \overset{(1)}{=} y - x - 2\varepsilon = |x - y| - 2\varepsilon.$$

Hence $V_\varepsilon(f, I) \leq V(g_\varepsilon, I) = |x-y|-2\varepsilon$. This proves the upper line in (3.37).

Case (II) Here we rely on the full form of (2.9). Let a sequence $\{\varepsilon_k\}_{k=1}^\infty$ be such that $0 < \varepsilon_k < \varepsilon = \tfrac{1}{2}|x - y|$ for all $k \in \mathbb{N}$ and $\varepsilon_k \to \varepsilon$ as $k \to \infty$. We set $g_k = g_{\varepsilon_k}$, $k \in \mathbb{N}$, where g_{ε_k} is defined in (3.38) (with $\varepsilon = \varepsilon_k$). By Case (I), given $k \in \mathbb{N}$, $g_k \in BV(I; M)$, $V(g_k, I) = |x - y| - 2\varepsilon_k$, and $d_{\infty,I}(f, g_k) = \varepsilon_k < \varepsilon$. Since $V(g_k, I) \to 0$ as $k \to \infty$, we conclude from (2.9) that $V_\varepsilon(f, I) = 0$.

Case (III) We set $c(t) = \varepsilon + \min\{x, y\}$, $t \in I$, and argue as in Example 3.6(e) (in Case (II) for $r = 0$). This gives $V_\varepsilon(f, I) = 0$, and completes the proof of (3.37).

Clearly, the metric space (M, d) in this example is *not proper*. Let us show that Lemma 2.14(b) is false. In fact, assume, for contradiction, that there is $g \in BV(I; M)$ with $d_{\infty,I}(f, g) \leq \varepsilon = \tfrac{1}{2}|x - y|$ such that $V_\varepsilon(f, I) = V(g, I)$. By (3.37), $V(g, I) = 0$, and so, $g = c$ is a constant function $c : I \to M$. From $d_{\infty,I}(f, c) \leq \varepsilon$, we find $|x - c| = |f(\tau) - c| \leq \varepsilon$, and so (as above, $x < y$),

$$c \leq x + \varepsilon = x + \tfrac{1}{2}(y - x) = \tfrac{1}{2}(x + y),$$

and, if $t \neq \tau$, then $|y - c| = |f(t) - c| \leq \varepsilon$, which implies

$$c \geq y - \varepsilon = y - \tfrac{1}{2}(y - x) = \tfrac{1}{2}(x + y).$$

Hence, $c = c(t) = \tfrac{1}{2}(x + y)$, $t \in I$, but $g = c \notin M^I$, which is a contradiction.

Example 3.11 Here we show that the assumption that the metric space (M, d) is *proper* in Lemma 2.14(c) is essential.

Let $x, y \in \mathbb{R}$, $x \neq y$, $M = \mathbb{R} \setminus \{\tfrac{1}{2}(x + y)\}$ with metric $d(u, v) = |u - v|$, $u, v \in M$, $I = [0, 1]$, and the sequence $\{f_j\} \subset M^I$ be given by

$$f_j(t) = \begin{cases} x & \text{if } j!t \text{ is an integer,} \\ y & \text{otherwise,} \end{cases} \quad t \in I, \ j \in \mathbb{N}. \tag{3.39}$$

We claim that, for all $j \in \mathbb{N}$,

$$V_\varepsilon(f_j, I) = \begin{cases} 2 \cdot j!\,(|x - y| - 2\varepsilon) & \text{if } 0 < \varepsilon < \tfrac{1}{2}|x - y|, \\ 0 & \text{if } \varepsilon \geq \tfrac{1}{2}|x - y|. \end{cases} \tag{3.40}$$

Suppose that we have already established (3.40). The sequence $\{f_j\}$ from (3.39) converges pointwise on I to the Dirichlet function $f = \mathscr{D}_{x,y}$ from (3.28). Let $\varepsilon = \tfrac{1}{2}|x - y|$. By (3.35) with $r = 0$, we have $V_\varepsilon(f, I) = \infty$, while, by (3.40), we get $V_\varepsilon(f_j, I) = 0$ for all $j \in \mathbb{N}$, and so, $\lim_{j \to \infty} V_\varepsilon(f_j, I) = 0$. Thus, the properness of metric space (M, d) in Lemma 2.14(c) is indispensable.

Proof of (3.40) In what follows, we fix $j \in \mathbb{N}$. By (2.11), $V_\varepsilon(f_j, I) = 0$ for all $\varepsilon \geq |f_j(I)| = |x - y|$. Now, we consider cases (I)–(III) from Example 3.10.

Case (I) Setting $t_k = k/j!$, we find from (3.39) that

$$f_j(t_k) = x \quad \text{for} \quad k = 0, 1, \ldots, j!,$$

and setting $s_k = \tfrac{1}{2}(t_{k-1} + t_k) = (k - \tfrac{1}{2})/j!$, we get

$$f_j(s_k) = y \quad \text{for} \quad k = 1, 2, \ldots, j!.$$

So, we have the following partition of the interval $I = [0, 1]$:

$$0 = t_0 < s_1 < t_1 < s_2 < t_2 < \ldots < s_{j!-1} < t_{j!-1} < s_{j!} < t_{j!} = 1. \tag{3.41}$$

If $g \in M^I$ is arbitrary with $d_{\infty,I}(f_j, g) \le \varepsilon$, then, applying (2.2), we obtain

$$V(g, I) \ge \sum_{k=1}^{j!} (|g(t_k) - g(s_k)| + |g(s_k) - g(t_{k-1})|)$$

$$\ge \sum_{k=1}^{j!} (|f_j(t_k) - f_j(s_k)| - 2\varepsilon + |f_j(s_k) - f_j(t_{k-1})| - 2\varepsilon)$$

$$= 2 \cdot j! (|x - y| - 2\varepsilon),$$

and so, by definition (2.6), $V_\varepsilon(f_j, I) \ge 2 \cdot j! (|x - y| - 2\varepsilon)$.

Now, we define a test function g_ε on I by (cf. (3.13)): given $t \in I$,

$$g_\varepsilon(t) = x - \varepsilon e_{x,y} \text{ if } j!t \text{ is an integer, and } g_\varepsilon(t) = y + \varepsilon e_{x,y} \text{ otherwise,}$$

$$\tag{3.42}$$

where $e_{x,y} = (x - y)/|x - y|$. Due to the symmetry in x and y, we may assume that $x < y$, and so, $g_\varepsilon(t) = x + \varepsilon$ if $j!t$ is an integer, and $g_\varepsilon(t) = y - \varepsilon$ otherwise, $t \in I$. We first note that $g_\varepsilon : I \to M$; in fact, if $j!t$ is an integer, then

$$g_\varepsilon(t) = x + \varepsilon < x + \tfrac{1}{2}|x - y| = x + \tfrac{1}{2}(y - x) = \tfrac{1}{2}(x + y),$$

and if $j!t$ is not an integer, then

$$g_\varepsilon(t) = y - \varepsilon > y - \tfrac{1}{2}|x - y| = y - \tfrac{1}{2}(y - x) = \tfrac{1}{2}(x + y).$$

Clearly, $d_{\infty,I}(f_j, g_\varepsilon) = \varepsilon$ and, by the additivity property (V.2) of V, for the partition (3.41), we find

$$V(g_\varepsilon, I) = \sum_{k=1}^{j!} (V(g_\varepsilon, [t_{k-1}, s_k]) + V(g_\varepsilon, [s_k, t_k]))$$

$$= \sum_{k=1}^{j!} (|g_\varepsilon(s_k) - g_\varepsilon(t_{k-1})| + |g_\varepsilon(t_k) - g_\varepsilon(s_k)|)$$

$$= \sum_{k=1}^{j!} (|(y - \varepsilon) - (x + \varepsilon)| + |(x + \varepsilon) - (y - \varepsilon)|)$$

$$= 2 \cdot j! (|x - y| - 2\varepsilon).$$

Thus, $V_\varepsilon(f_j, I) \le V(g_\varepsilon, I)$, and this implies the upper line in (3.40).

Case (II) Let a sequence $\{\varepsilon_k\}_{k=1}^{\infty}$ be such that $0 < \varepsilon_k < \varepsilon = \frac{1}{2}|x - y|$, $k \in \mathbb{N}$, and $\varepsilon_k \to \varepsilon$ as $k \to \infty$. Set $g_k = g_{\varepsilon_k}$, $k \in \mathbb{N}$, where g_{ε_k} is given by (3.42) (with $x < y$). We know from Case (I) that, for every $k \in \mathbb{N}$, $g_k \in \mathrm{BV}(I; M)$,

$$V(g_k, I) = 2 \cdot j! \, (|x - y| - 2\varepsilon_k), \quad \text{and} \quad d_{\infty, I}(f_j, g_k) = \varepsilon_k < \varepsilon.$$

Since $V(g_k, I) \to 0$ as $k \to \infty$, we conclude from (2.9) that $V_\varepsilon(f_j, I) = 0$.

Case (III) We set $c = c(t) = \varepsilon + \min\{x, y\}$ for all $t \in I$, i.e., under our assumption $x < y$, $c = \varepsilon + x$. Note that $c \in M$, because assumptions (III) and $x < y$ imply $\varepsilon > \frac{1}{2}(y - x)$, and so, $c = \varepsilon + x > \frac{1}{2}(y - x) + x = \frac{1}{2}(x + y)$. Furthermore, $d_{\infty, I}(f_j, c) \le \varepsilon$; in fact, given $t \in I$, if $j!t$ is an integer, then $|f_j(t) - c(t)| = |x - c| = \varepsilon$, and if $j!t$ is not an integer, then

$$|f_j(t) - c(t)| = |y - x - \varepsilon| \overset{\text{(III)}}{=} y - x - \varepsilon < |x - y| - \tfrac{1}{2}|x - y| = \tfrac{1}{2}|x - y| < \varepsilon.$$

Since c is a constant function from M^I, we get $V_\varepsilon(f_j, I) = 0$. This completes the proof of (3.40).

Chapter 4
Pointwise selection principles

4.1 Functions with values in a metric space

Our first main result, an extension of Theorem 3.8 from [43], is a *pointwise selection principle* for metric space valued univariate functions in terms of the approximate variation (see Theorem 4.1).

In order to formulate it, we slightly generalize the notion of a regulated function (cf. p. 9). If $T \subset \mathbb{R}$ is an arbitrary set and (M, d) is a metric space, a function $f \in M^T$ is said to be *regulated* on T (in symbols, $f \in \mathrm{Reg}(T; M)$) if it satisfies the Cauchy condition at every left limit point of T and every right limit point of T. More explicitly, given $\tau \in T$, which is a *left limit point* of T (i.e., $T \cap (\tau - \delta, \tau) \neq \varnothing$ for all $\delta > 0$), we have $d(f(s), f(t)) \to 0$ as $T \ni s, t \to \tau - 0$; and given $\tau' \in T$, which is a *right limit point* of T (i.e., $T \cap (\tau', \tau' + \delta) \neq \varnothing$ for all $\delta > 0$), we have $d(f(s), f(t)) \to 0$ as $T \ni s, t \to \tau' + 0$. The proof of Lemma 2.10 in ($\supset$) shows that

$$\mathrm{Reg}(T; M) \supset \{f \in M^T : V_\varepsilon(f, T) < \infty \text{ for all } \varepsilon > 0\}; \tag{4.1}$$

in fact, it suffices to set $\varphi_\varepsilon(t) = V_\varepsilon(f, T \cap (-\infty, t])$, $t \in T$, and treat s, t from T.

In contrast to the case when $T = I$ is an interval (see p. 9), a function f from $\mathrm{Reg}(T; M)$ may not be bounded in general: for instance, $f \in \mathbb{R}^T$ given on the set $T = [0, 1] \cup \{2 - \frac{1}{n}\}_{n=2}^\infty$ by: $f(t) = t$ if $0 \le t \le 1$ and $f(2 - \frac{1}{n}) = n$ if $n \in \mathbb{N}$ and $n \ge 2$, is regulated in the above sense, but not bounded.

In what follows, we denote by $\mathrm{Mon}(T; \mathbb{R}^+)$ the set of all bounded nondecreasing functions mapping T into $\mathbb{R}^+ = [0, \infty)$ ($\mathbb{R}^+$ may be replaced by $\mathbb{R}$). It is worthwhile to recall the classical *Helly selection principle* for an arbitrary set $T \subset \mathbb{R}$ (e.g., [18, Proof of Theorem 1.3]): *a uniformly bounded sequence of functions from $\mathrm{Mon}(T; \mathbb{R})$ contains a subsequence which converges pointwise on T to a function from $\mathrm{Mon}(T; \mathbb{R})$.*

© The Author(s), under exclusive license to Springer Nature Switzerland AG 2021
V. V. Chistyakov, *From Approximate Variation to Pointwise Selection Principles*,
SpringerBriefs in Optimization, https://doi.org/10.1007/978-3-030-87399-8_4

Theorem 4.1 *Let $\varnothing \neq T \subset \mathbb{R}$ and (M, d) be a metric space. If $\{f_j\} \subset M^T$ is a pointwise relatively compact sequence of functions on T such that*

$$\limsup_{j \to \infty} V_\varepsilon(f_j, T) < \infty \quad \text{for all} \quad \varepsilon > 0, \tag{4.2}$$

then there is a subsequence of $\{f_j\}$, which converges pointwise on T to a bounded regulated function $f \in M^T$. In addition, if (M, d) is proper metric space, then $V_\varepsilon(f, T)$ does not exceed the $\limsup$ in (4.2) for all $\varepsilon > 0$.

Proof We present a direct proof based only on the properties of the approximate variation from Sect. 2.4 (an indirect proof, based on the notion of the *joint modulus of variation of two functions*, was given in [29, Theorem 3]).

By Lemma 2.2(b), given $\varepsilon > 0$ and $j \in \mathbb{N}$, the ε-variation function defined by the rule $t \mapsto V_\varepsilon(f_j, T \cap (-\infty, t])$ is nondecreasing on T. Note also that, by assumption (4.2), for each $\varepsilon > 0$ there are $j_0(\varepsilon) \in \mathbb{N}$ and a number $C(\varepsilon) > 0$ such that

$$V_\varepsilon(f_j, T) \leq C(\varepsilon) \quad \text{for all} \quad j \geq j_0(\varepsilon).$$

We divide the rest of the proof into five steps.

1. Let us show that for each decreasing sequence $\{\varepsilon_k\}_{k=1}^\infty$ of positive numbers $\varepsilon_k \to 0$ there are a subsequence of $\{f_j\}$, again denoted by $\{f_j\}$, and a sequence of functions $\{\varphi_k\}_{k=1}^\infty \subset \mathrm{Mon}(T; \mathbb{R}^+)$ such that

$$\lim_{j \to \infty} V_{\varepsilon_k}(f_j, T \cap (-\infty, t]) = \varphi_k(t) \quad \text{for all } k \in \mathbb{N} \text{ and } t \in T. \tag{4.3}$$

In order to prove (4.3), we make use of Cantor's diagonal procedure. Lemma 2.2(b) and remarks above imply

$$V_{\varepsilon_1}(f_j, T \cap (-\infty, t]) \leq V_{\varepsilon_1}(f_j, T) \leq C(\varepsilon_1) \text{ for all } t \in T \text{ and } j \geq j_0(\varepsilon_1),$$

i.e., the sequence of functions $\{t \mapsto V_{\varepsilon_1}(f_j, T \cap (-\infty, t])\}_{j=j_0(\varepsilon_1)}^\infty \subset \mathrm{Mon}(T; \mathbb{R}^+)$ is uniformly bounded on T by constant $C(\varepsilon_1)$. By the classical Helly selection principle (for monotone functions), there are a subsequence $\{J_1(j)\}_{j=1}^\infty$ of $\{j\}_{j=j_0(\varepsilon_1)}^\infty$ and a function $\varphi_1 \in \mathrm{Mon}(T; \mathbb{R}^+)$ such that $V_{\varepsilon_1}(f_{J_1(j)}, T \cap (-\infty, t])$ converges to $\varphi_1(t)$ in $\mathbb{R}$ as $j \to \infty$ for all $t \in T$. Now, choose the least number $j_1 \in \mathbb{N}$ such that $J_1(j_1) \geq j_0(\varepsilon_2)$. Inductively, assume that $k \in \mathbb{N}$, $k \geq 2$, and a subsequence $\{J_{k-1}(j)\}_{j=1}^\infty$ of $\{j\}_{j=j_0(\varepsilon_1)}^\infty$ and a number $j_{k-1} \in \mathbb{N}$ with $J_{k-1}(j_{k-1}) \geq j_0(\varepsilon_k)$ are already constructed. By Lemma 2.2(b), we get

$$V_{\varepsilon_k}(f_{J_{k-1}(j)}, T \cap (-\infty, t]) \leq V_{\varepsilon_k}(f_{J_{k-1}(j)}, T) \leq C(\varepsilon_k) \text{ for all } t \in T \text{ and } j \geq j_{k-1},$$

and so, by the Helly selection principle, there are a subsequence $\{J_k(j)\}_{j=1}^\infty$ of the sequence $\{J_{k-1}(j)\}_{j=j_{k-1}}^\infty$ and a function $\varphi_k \in \mathrm{Mon}(T; \mathbb{R}^+)$ such that

$$\lim_{j \to \infty} V_{\varepsilon_k}(f_{J_k(j)}, T \cap (-\infty, t]) = \varphi_k(t) \quad \text{for all} \quad t \in T.$$

Given $k \in \mathbb{N}$, $\{J_j(j)\}_{j=k}^\infty$ is a subsequence of $\{J_k(j)\}_{j=1}^\infty$, and so, the diagonal sequence $\{f_{J_j(j)}\}_{j=1}^\infty$, again denoted by $\{f_j\}$, satisfies condition (4.3).

2. Let Q be an at most countable dense subset of T. Note that any point $t \in T$, which is not a limit point for T (i.e., $T \cap (t - \delta, t + \delta) = \{t\}$ for some $\delta > 0$), belongs to Q. Since, for any $k \in \mathbb{N}$, $\varphi_k \in \mathrm{Mon}(T; \mathbb{R}^+)$, the set $Q_k \subset T$ of points of discontinuity of φ_k is at most countable. Setting $S = Q \cup \bigcup_{k=1}^\infty Q_k$, we find that S is an at most countable dense subset of T; moreover, if $S \neq T$, then every point $t \in T \setminus S$ is a limit point for T and

$$\varphi_k \text{ is continuous on } T \setminus S \text{ for all } k \in \mathbb{N}. \tag{4.4}$$

Since $S \subset T$ is at most countable and $\{f_j(s) : j \in \mathbb{N}\}$ is relatively compact in M for all $s \in S$, applying the Cantor diagonal procedure and passing to a subsequence of $\{f_j(s)\}_{j=1}^\infty$ if necessary, with no loss of generality we may assume that, for each $s \in S$, $f_j(s)$ converges in M as $j \to \infty$ to a (unique) point denoted by $f(s) \in M$ (so that $f : S \to M$).

If $S = T$, we turn to Step 4 below and complete the proof.

3. Now, assuming that $S \neq T$, we prove that $f_j(t)$ converges in M as $j \to \infty$ for all $t \in T \setminus S$, as well. Let $t \in T \setminus S$ and $\eta > 0$ be arbitrarily fixed. Since $\varepsilon_k \to 0$ as $k \to \infty$ (cf. Step 1), we pick and fix $k = k(\eta) \in \mathbb{N}$ such that $\varepsilon_k \leq \eta$. By (4.4), φ_k is continuous at t, and so, by the density of S in T, there is $s = s(k, t) \in S$ such that $|\varphi_k(t) - \varphi_k(s)| \leq \eta$. From property (4.3), there is $j^1 = j^1(\eta, k, t, s) \in \mathbb{N}$ such that, for all $j \geq j^1$,

$$|V_{\varepsilon_k}(f_j, T \cap (-\infty, t]) - \varphi_k(t)| \leq \eta \quad \text{and} \quad |V_{\varepsilon_k}(f_j, T \cap (-\infty, s]) - \varphi_k(s)| \leq \eta. \tag{4.5}$$

Assuming that $s < t$ (with no loss of generality) and applying Lemma 2.8 (where T is replaced by $T \cap (-\infty, t]$, T_1—by $T \cap (-\infty, s]$, and T_2—by $T \cap [s, t]$), we get

$$V_{\varepsilon_k}(f_j, T \cap [s, t]) \leq V_{\varepsilon_k}(f_j, T \cap (-\infty, t]) - V_{\varepsilon_k}(f_j, T \cap (-\infty, s])$$

$$\leq |V_{\varepsilon_k}(f_j, T \cap (-\infty, t]) - \varphi_k(t)| + |\varphi_k(t) - \varphi_k(s)|$$

$$+ |\varphi_k(s) - V_{\varepsilon_k}(f_j, T \cap (-\infty, s])|$$

$$\leq \eta + \eta + \eta = 3\eta \quad \text{for all} \quad j \geq j^1.$$

By the definition of $V_{\varepsilon_k}(f_j, T \cap [s, t])$, for each $j \geq j^1$, there is $g_j \in \mathrm{BV}(T \cap [s, t]; M)$ (also depending on η, k, t, and s) such that

$$d_{\infty, T \cap [s,t]}(f_j, g_j) \leq \varepsilon_k \quad \text{and} \quad V(g_j, T \cap [s, t]) \leq V_{\varepsilon_k}(f_j, T \cap [s, t]) + \eta.$$

These inequalities, (2.2), and property (**V.1**) from p. 8 yield, for all $j \geq j^1$,

$$d(f_j(s), f_j(t)) \leq d(g_j(s), g_j(t)) + 2 d_{\infty, T \cap [s,t]}(f_j, g_j)$$

$$\leq V(g_j, T \cap [s, t]) + 2\varepsilon_k \leq (3\eta + \eta) + 2\eta = 6\eta. \quad (4.6)$$

Being convergent, the sequence $\{f_j(s)\}_{j=1}^{\infty}$ is Cauchy in M, and so, there is a natural number $j^2 = j^2(\eta, s)$ such that $d(f_j(s), f_{j'}(s)) \leq \eta$ for all $j, j' \geq j^2$. Since the number $j^3 = \max\{j^1, j^2\}$ depends only on η (and t) and

$$d(f_j(t), f_{j'}(t)) \leq d(f_j(t), f_j(s)) + d(f_j(s), f_{j'}(s)) + d(f_{j'}(s), f_{j'}(t))$$

$$\leq 6\eta + \eta + 6\eta = 13\eta \quad \text{for all} \quad j, j' \geq j^3,$$

the sequence $\{f_j(t)\}_{j=1}^{\infty}$ is Cauchy in M. Taking into account that $\{f_j(t) : j \in \mathbb{N}\}$ is relatively compact in M, we conclude that $f_j(t)$ converges in M as $j \to \infty$ to a (unique) point denoted by $f(t) \in M$ (so, $f : T \setminus S \to M$).

4. At the end of Steps 2 and 3, we have shown that the function f mapping $T = S \cup (T \setminus S)$ into M is the pointwise limit on T of a subsequence $\{f_{j_p}\}_{p=1}^{\infty}$ of the original sequence $\{f_j\}_{j=1}^{\infty}$. By virtue of Lemma 2.5(b) and assumption (4.2), given $\varepsilon > 0$, we get

$$|f(T)| \leq \liminf_{p \to \infty} |f_{j_p}(T)| \leq \liminf_{p \to \infty} V_{\varepsilon}(f_{j_p}, T) + 2\varepsilon$$

$$\leq \limsup_{j \to \infty} V_{\varepsilon}(f_j, T) + 2\varepsilon < \infty,$$

and so, f is a *bounded* function on T, i.e., $f \in \mathrm{B}(T; M)$.

Now, we prove that f is *regulated* on T. Given $\tau \in T$, which is a left limit point for T, let us show that $d(f(s), f(t)) \to 0$ as $T \ni s, t \to \tau - 0$ (similar arguments apply if $\tau' \in T$ is a right limit point for T). This is equivalent to showing that for every $\eta > 0$ there is $\delta = \delta(\eta) > 0$ such that $d(f(s), f(t)) \leq 7\eta$ for all $s, t \in T \cap (\tau - \delta, \tau)$ with $s < t$. Recall that the (finally) extracted subsequence of the original sequence $\{f_j\}$, here again denoted by $\{f_j\}$, satisfies condition (4.3) and $f_j \to f$ pointwise on T.

Let $\eta > 0$ be arbitrarily fixed. Since $\varepsilon_k \to 0$, pick and fix $k = k(\eta) \in \mathbb{N}$ such that $\varepsilon_k \leq \eta$. Furthermore, since $\varphi_k \in \mathrm{Mon}(T; \mathbb{R}^+)$ and $\tau \in T$ is a left limit point of T, the left limit $\lim_{T \ni t \to \tau - 0} \varphi_k(t) \in \mathbb{R}^+$ exists. Hence, there is $\delta = \delta(\eta, k) > 0$ such that $|\varphi_k(t) - \varphi_k(s)| \leq \eta$ for all $s, t \in T \cap (\tau - \delta, \tau)$. Now, let $s, t \in T \cap (\tau - \delta, \tau)$ be arbitrary. By (4.3), there is $j^1 = j^1(\eta, k, s, t) \in \mathbb{N}$

such that if $j \geq j^1$, the inequalities (4.5) hold. Arguing exactly the same way as between lines (4.5) and (4.6), we find that $d(f_j(s), f_j(t)) \leq 6\eta$ for all $j \geq j^1$. Noting that $f_j(s) \to f(s)$ and $f_j(t) \to f(t)$ in M as $j \to \infty$, by the triangle inequality for d, we have, as $j \to \infty$,

$$|d(f_j(s), f_j(t)) - d(f(s), f(t))| \leq d(f_j(s), f(s)) + d(f_j(t), f(t)) \to 0.$$

So, there is $j^2 = j^2(\eta, s, t) \in \mathbb{N}$ such that $d(f(s), f(t)) \leq d(f_j(s), f_j(t)) + \eta$ for all $j \geq j^2$. Thus, if $j \geq \max\{j^1, j^2\}$, we get $d(f(s), f(t)) \leq 6\eta + \eta = 7\eta$.

5. Finally, assume that (M, d) is a *proper* metric space. Once again (as at the beginning of Step 4) it is convenient to denote the pointwise convergent subsequence of $\{f_j\}$ by $\{f_{j_p}\}_{p=1}^{\infty}$. So, since $f_{j_p} \to f$ pointwise on T as $p \to \infty$, we may apply Lemma 2.14(c) and assumption (4.2) and get, for all $\varepsilon > 0$,

$$V_\varepsilon(f, T) \leq \liminf_{p \to \infty} V_\varepsilon(f_{j_p}, T) \leq \limsup_{j \to \infty} V_\varepsilon(f, T) < \infty.$$

This and (4.1) (or Lemma 2.10 if $T = I$) also imply $f \in \text{Reg}(T; M)$.

This completes the proof of Theorem 4.1. $\qquad\qquad\square$

A few remarks concerning Theorem 4.1 are in order (see also Remarks 4.5 and 4.6 below).

Remark 4.2 If (M, d) is a *proper* metric space, then the assumption "$\{f_j\} \subset M^T$ is pointwise relatively compact on T" in Theorem 4.1 can be replaced by an (seemingly weaker, but, actually) equivalent condition "$\{f_j\} \subset M^T$ and $\{f_j(t_0)\}$ is *eventually bounded* in M for some $t_0 \in T$" in the sense that there are $J_0 \in \mathbb{N}$ and a constant $C_0 > 0$ such that $d(f_j(t_0), f_{j'}(t_0)) \leq C_0$ for all $j, j' \geq J_0$. In fact, fixing $\varepsilon > 0$, e.g., $\varepsilon = 1$, by Lemma 2.5(b), we get $|f_j(T)| \leq V_1(f_j, T) + 2$ for all $j \in \mathbb{N}$, and so, applying assumption (4.2),

$$\limsup_{j \to \infty} |f_j(T)| \leq \limsup_{j \to \infty} V_1(f_j, T) + 2 < \infty.$$

Hence, there are $J_1 \in \mathbb{N}$ and a constant $C_1 > 0$ such that $|f_j(T)| \leq C_1$ for all $j \geq J_1$. By the triangle inequality for d, given $t \in T$, we find, for all $j, j' \geq \max\{J_0, J_1\}$,

$$d(f_j(t), f_{j'}(t)) \leq d(f_j(t), f_j(t_0)) + d(f_j(t_0), f_{j'}(t_0)) + d(f_{j'}(t_0), f_{j'}(t))$$

$$\leq |f_j(T)| + C_0 + |f_{j'}(T)| \leq C_1 + C_0 + C_1, \tag{4.7}$$

i.e., $\{f_j(t)\}$ is eventually bounded uniformly in $t \in T$. Thus, since M is proper, $\{f_j(t)\}$ is relatively compact in M for all $t \in T$. In the case under consideration, an alternative proof of Theorem 4.1 can be given (see Theorem 4.3 and its proof).

However, for a general metric space (M, d), the relative compactness of $\{f_j(t)\}$ at all points $t \in T$ cannot be replaced by their (closedness and) boundedness even at a single point of T. To see this, let $T = I = [a, b]$ and $M = \ell^1 \subset \mathbb{R}^{\mathbb{N}}$ be the (infinite-dimensional) Banach space of all summable sequences $u = \{u_n\}_{n=1}^{\infty} \in \ell^1$ equipped with the norm $\|u\| = \sum_{n=1}^{\infty} |u_n| < \infty$. If $j \in \mathbb{N}$, denote by $e_j = \{u_n\}_{n=1}^{\infty}$ the unit vector from ℓ^1 given by $u_n = 0$ if $n \neq j$, and $u_j = 1$. Now, define the sequence $\{f_j\} \subset M^T$ by $f_j(a) = e_j$ and $f_j(t) = 0$ if $a < t \leq b$, $j \in \mathbb{N}$. We have: the set $\{f_j(a)\}_{j=1}^{\infty} = \{e_j : j \in \mathbb{N}\}$ is closed and bounded in M, $\{f_j(t)\}_{j=1}^{\infty} = \{0\}$ is compact in M if $a < t \leq b$, and (cf. Example 3.3 and (3.10)), given $j \in \mathbb{N}$, $V_\varepsilon(f_j, T) = 1 - 2\varepsilon$ if $0 < \varepsilon < 1/2$, and $V_\varepsilon(f_j, T) = 0$ if $\varepsilon \geq 1/2$. Clearly, condition (4.2) is satisfied for $\{f_j\}$, but no subsequence of $\{f_j\}$ converges in M at the point $t = a$.

Theorem 4.3 *Suppose $T \subset \mathbb{R}$, (M, d) is a proper metric space, and a sequence of functions $\{f_j\} \subset M^T$ is such that $\{f_j(t_0)\}$ is eventually bounded in M for some $t_0 \in T$, and condition (4.2) holds. Then, a subsequence of $\{f_j\}$ converges pointwise on T to a bounded function $f \in M^T$ such that*

$$V_\varepsilon(f, T) \leq \limsup_{j \to \infty} V_\varepsilon(f_j, T) \quad \text{for all} \quad \varepsilon > 0.$$

Proof

1. Let $\{\varepsilon_k\}_{k=1}^{\infty} \subset (0, \infty)$ be such that $\varepsilon_k \to 0$ as $k \to \infty$. Given $k \in \mathbb{N}$, condition (4.2) implies the existence of $j_0'(\varepsilon_k) \in \mathbb{N}$ and a constant $C(\varepsilon_k) > 0$ such that $V_{\varepsilon_k}(f_j, T) < C(\varepsilon_k)$ for all $j \geq j_0'(\varepsilon_k)$. By definition (2.6), for each $j \geq j_0'(\varepsilon_k)$, there is $g_j^{(k)} \in BV(T; M)$ such that[1]

$$d_{\infty,T}(f_j, g_j^{(k)}) \leq \varepsilon_k \quad \text{and} \quad V(g_j^{(k)}, T) \leq C(\varepsilon_k). \quad (4.8)$$

Since $\{f_j(t_0)\}$ is eventually bounded and (4.2) holds, we get inequality (4.7) for all $j, j' \geq \max\{J_0, J_1\}$. It follows that if $t \in T$, $k \in \mathbb{N}$, and $j, j' \geq j_0(\varepsilon_k) \equiv \max\{j_0'(\varepsilon_k), J_0, J_1\}$, we find, by the triangle inequality for d, (4.8), and (4.7),

$$d(g_j^{(k)}(t), g_{j'}^{(k)}(t)) \leq d(g_j^{(k)}(t), f_j(t)) + d(f_j(t), f_{j'}(t)) + d(f_{j'}(t), g_{j'}^{(k)}(t))$$

$$\leq d_{\infty,T}(g_j^{(k)}, f_j) + d(f_j(t), f_{j'}(t)) + d_{\infty,T}(f_{j'}, g_{j'}^{(k)})$$

$$\leq \varepsilon_k + (C_0 + 2C_1) + \varepsilon_k.$$

---

[1] Conditions "$\{f_j\} \subset M^T$ is relatively compact on T and $\{g_j\} \subset M^T$ is such that $d_{\infty,T}(f_j, g_j) \leq \varepsilon$ for all $j \in \mathbb{N}$" *do not* imply in general that $\{g_j\}$ is also relatively compact on T: for instance (cf. notation in Remark 4.2), $T = [0, 1]$, $M = \ell^1$ (which is not proper), $f_j(t) = 0$ and $g_j(t) = \varepsilon t e_j$ for all $j \in \mathbb{N}$ and $t \in T$.

In this way, we have shown that

$$\sup_{j,j' \geq j_0(\varepsilon_k)} d(g_j^{(k)}(t), g_{j'}^{(k)}(t)) \leq 2\varepsilon_k + C_0 + 2C_1 \quad \text{for all } k \in \mathbb{N} \text{ and } t \in T,$$
(4.9)

and, by the second inequality in (4.8),

$$\sup_{j \geq j_0(\varepsilon_k)} V(g_j^{(k)}, T) \leq C(\varepsilon_k) \quad \text{for all } k \in \mathbb{N}.$$
(4.10)

2. Applying Cantor's diagonal procedure, let us show the following: given $k \in \mathbb{N}$, there exist a subsequence of $\{g_j^{(k)}\}_{j=j_0(\varepsilon_k)}^{\infty}$, denoted by $\{g_j^{(k)}\}_{j=1}^{\infty}$, and $g^{(k)} \in$ BV$(T; M)$ such that

$$\lim_{j \to \infty} d(g_j^{(k)}(t), g^{(k)}(t)) = 0 \quad \text{for all } t \in T.$$
(4.11)

Setting $k = 1$ in (4.9) and (4.10), we find that the sequence $\{g_j^{(1)}\}_{j=j_0(\varepsilon_1)}^{\infty}$ has uniformly bounded (by $C(\varepsilon_1)$) Jordan variations and is uniformly bounded on T (by $2\varepsilon_1 + C_0 + 2C_1$), and so, since M is a *proper* metric space, the sequence is pointwise relatively compact on T. By the Helly-type pointwise selection principle in BV$(T; M)$ (cf. property (**V.4**) on p. 9), there are a subsequence $\{J_1(j)\}_{j=1}^{\infty}$ of $\{j\}_{j=j_0(\varepsilon_1)}^{\infty}$ and a function $g^{(1)} \in$ BV$(T; M)$ such that $g_{J_1(j)}^{(1)}(t) \to g^{(1)}(t)$ in M as $j \to \infty$ for all $t \in T$. Pick the least number $j_1 \in \mathbb{N}$ such that $J_1(j_1) \geq j_0(\varepsilon_2)$. Inductively, if $k \in \mathbb{N}$ with $k \geq 2$, a subsequence $\{J_{k-1}(j)\}_{j=1}^{\infty}$ of $\{j\}_{j=j_0(\varepsilon_1)}^{\infty}$, and a number $j_{k-1} \in \mathbb{N}$ such that $J_{k-1}(j_{k-1}) \geq j_0(\varepsilon_k)$ are already chosen, we get the sequence of functions $\{g_{J_{k-1}(j)}^{(k)}\}_{j=j_{k-1}}^{\infty} \subset$ BV$(T; M)$, which, by virtue of (4.9) and (4.10), satisfies conditions:

$$\sup_{j,j' \geq j_{k-1}} d\left(g_{J_{k-1}(j)}^{(k)}(t), g_{J_{k-1}(j')}^{(k)}(t)\right) \leq 2\varepsilon_k + C_0 + 2C_1 \quad \text{for all } t \in T$$

and

$$\sup_{j \geq j_{k-1}} V\left(g_{J_{k-1}(j)}^{(k)}, T\right) \leq C(\varepsilon_k).$$

By Helly's-type selection principle (**V.4**) in BV$(T; M)$, there are a subsequence $\{J_k(j)\}_{j=1}^{\infty}$ of $\{J_{k-1}(j)\}_{j=j_{k-1}}^{\infty}$ and a function $g^{(k)} \in$ BV$(T; M)$ such that $g_{J_k(j)}^{(k)}(t) \to g^{(k)}(t)$ in M as $j \to \infty$ for all $t \in T$. Since, for each $k \in \mathbb{N}$, $\{J_j(j)\}_{j=k}^{\infty}$ is a subsequence of $\{J_k(j)\}_{j=j_{k-1}}^{\infty} \subset \{J_k(j)\}_{j=1}^{\infty}$, we conclude that the diagonal sequence $\{g_{J_j(j)}^{(k)}\}_{j=1}^{\infty}$, (which was) denoted by $\{g_j^{(k)}\}_{j=1}^{\infty}$ (at the beginning of step 2), satisfies condition (4.11).

We denote the corresponding diagonal subsequence $\{f_{J_j(j)}\}_{j=1}^\infty$ of $\{f_j\}$ again by $\{f_j\}$.

3. Since $BV(T; M) \subset B(T; M)$ (by (V.1) on p. 8), $\{g^{(k)}\}_{k=1}^\infty \subset B(T; M)$. We are going to show that $\{g^{(k)}\}_{k=1}^\infty$ is a Cauchy sequence with respect to the uniform metric $d_{\infty,T}$. For this, we employ an idea from [43, p. 49].

Let $\eta > 0$ be arbitrary. From $\varepsilon_k \to 0$, we find $k_0 = k_0(\eta) \in \mathbb{N}$ such that $\varepsilon_k \leq \eta$ for all $k \geq k_0$. Now, suppose $k, k' \in \mathbb{N}$ are (arbitrary) such that $k, k' \geq k_0$. By virtue of (4.11), for each $t \in T$, there is a number $j^1 = j^1(t, \eta, k, k') \in \mathbb{N}$ such that if $j \geq j^1$, we have

$$d\big(g_j^{(k)}(t), g^{(k)}(t)\big) \leq \eta \quad \text{and} \quad d\big(g_j^{(k')}(t), g^{(k')}(t)\big) \leq \eta.$$

It follows from the triangle inequality for d and the first inequality in (4.8) that if $j \geq j^1$,

$$d\big(g^{(k)}(t), g^{(k')}(t)\big) \leq d\big(g^{(k)}(t), g_j^{(k)}(t)\big) + d\big(g_j^{(k)}(t), f_j(t)\big)$$

$$+ d\big(f_j(t), g_j^{(k')}(t)\big) + d\big(g_j^{(k')}(t), g^{(k')}(t)\big)$$

$$\leq \eta + \varepsilon_k + \varepsilon_{k'} + \eta \leq 4\eta.$$

By the arbitrariness of $t \in T$, $d_{\infty,T}(g^{(k)}, g^{(k')}) \leq 4\eta$ for all $k, k' \geq k_0$.

4. Being proper, (M, d) is complete (cf. p. 21), and so, $B(T; M)$ is complete with respect to the uniform metric $d_{\infty,T}$. By step 3, there is $g \in B(T; M)$ such that $g^{(k)} \rightrightarrows g$ on T (i.e., $d_{\infty,T}(g^{(k)}, g) \to 0$) as $k \to \infty$. Let us prove that $f_j \to g$ pointwise on T as $j \to \infty$ ($\{f_j\}$ being from the end of step 2).

Let $t \in T$ and $\eta > 0$ be arbitrary. Choose and fix a number $k = k(\eta) \in \mathbb{N}$ such that $\varepsilon_k \leq \eta$ and $d_{\infty,T}(g^{(k)}, g) \leq \eta$. By (4.11), there is $j^2 = j^2(t, \eta, k) \in \mathbb{N}$ such that $d(g_j^{(k)}(t), g^{(k)}(t)) \leq \eta$ for all $j \geq j^2$, and so, (4.8) implies

$$d(f_j(t), g(t)) \leq d\big(f_j(t), g_j^{(k)}(t)\big) + d\big(g_j^{(k)}(t), g^{(k)}(t)\big) + d\big(g^{(k)}(t), g(t)\big)$$

$$\leq \varepsilon_k + d\big(g_j^{(k)}(t), g^{(k)}(t)\big) + d_{\infty,T}(g^{(k)}, g)$$

$$\leq \eta + \eta + \eta = 3\eta \quad \text{for all} \quad j \geq j^2,$$

which proves our assertion.

Thus, we have shown that a suitable (diagonal) subsequence $\{f_{j_p}\}_{p=1}^\infty$ of the original sequence $\{f_j\}_{j=1}^\infty$ converges pointwise on T to the function g from $B(T; M)$. Setting $f = g$ and applying Lemma 2.14(c), we conclude that

$$V_\varepsilon(f, T) = V_\varepsilon(g, T) \leq \liminf_{p \to \infty} V_\varepsilon(f_{j_p}, T) \leq \limsup_{j \to \infty} V_\varepsilon(f_j, T) \quad \forall \varepsilon > 0.$$

This completes the proof of Theorem 4.3. □

A simple consequence of Theorem 4.3 is the following.

Corollary 4.4 *Assume that assumption (4.2) in Theorem 4.3 is replaced by condition* $\lim_{j\to\infty} |f_j(T)| = 0$. *Then, a subsequence of* $\{f_j\}$ *converges pointwise on T to a constant function on T.*

Proof In fact, given $\varepsilon > 0$, there is $j_0 = j_0(\varepsilon) \in \mathbb{N}$ such that $|f_j(T)| \le \varepsilon$ for all $j \ge j_0$, and so, by (2.11), $V_\varepsilon(f_j, T) = 0$ for all $j \ge j_0$. This implies

$$\limsup_{j\to\infty} V_\varepsilon(f_j, T) \le \sup_{j \ge j_0} V_\varepsilon(f_j, T) = 0 \quad \text{for all} \quad \varepsilon > 0.$$

By Theorem 4.3, a subsequence of $\{f_j\}$ converges pointwise on T to a function $f \in M^T$ such that $V_\varepsilon(f, T) = 0$ for all $\varepsilon > 0$. Lemma 2.5(e) yields $|f(T)| = 0$, i.e., f is a constant function on T. □

Remark 4.5 The classical Helly selection principle for monotone functions (p. 45) is a particular case of Theorem 4.1. In fact, suppose $\{\varphi_j\} \subset \mathbb{R}^T$ is a sequence of monotone functions, for which there is a constant $C > 0$ such that $|\varphi_j(t)| \le C$ for all $t \in T$ and $j \in \mathbb{N}$. Setting $(M, \|\cdot\|) = (\mathbb{R}, |\cdot|)$, $x = 1$, and $y = 0$ in Example 3.1, for every $j \in \mathbb{N}$ we find, from (3.8) and (3.3), that $V_\varepsilon(\varphi_j, T) = |\varphi_j(T)| - 2\varepsilon$ whenever $0 < \varepsilon < \frac{1}{2}|\varphi_j(T)|$, and $V_\varepsilon(\varphi_j, T) = 0$ if $\varepsilon \ge \frac{1}{2}|\varphi_j(T)|$. Since $|\varphi_j(T)| \le 2C$, we get $V_\varepsilon(\varphi_j, T) \le 2C$ for all $j \in \mathbb{N}$ and $\varepsilon > 0$, so (4.2) is satisfied.

Similarly, Theorem 4.1 implies Helly's selection principle for functions of bounded variation (cf. property (**V.4**) on p. 9). In fact, if $\{f_j\} \subset M^T$ and $C = \sup_{j\in\mathbb{N}} V(f_j, T)$ is finite, then, by Lemma 2.5(a), $V_\varepsilon(f_j, T) \le C$ for all $j \in \mathbb{N}$ and $\varepsilon > 0$, so (4.2) is fulfilled. Now, if a subsequence of $\{f_j\}$ converges pointwise on T to $f \in M^T$, then property (**V.3**) (p. 9) implies $f \in \mathrm{BV}(T; M)$ with $V(f, T) \le C$.

Remark 4.6

(a) Condition (4.2) is *necessary* for the *uniform convergence* in the following sense: if $\{f_j\} \subset M^T$, $f_j \rightrightarrows f$ on T, and $V_\varepsilon(f, T) < \infty$ for all $\varepsilon > 0$, then, by Lemma 2.12(a),

$$\limsup_{j\to\infty} V_\varepsilon(f_j, T) \le V_{\varepsilon-0}(f, T) \le V_{\varepsilon'}(f, T) < \infty \quad \text{for all} \quad 0 < \varepsilon' < \varepsilon.$$

(b) Contrary to this, condition (4.2) *is not* necessary for the pointwise convergence (see Examples 4.9 and 4.11). On the other hand, condition (4.2) is "almost necessary" for the pointwise convergence $f_j \to f$ on T in the following sense. Assume that $T \subset \mathbb{R}$ is a measurable set with *finite* Lebesgue measure $\mathscr{L}(T)$ and $\{f_j\} \subset M^T$ is a sequence of measurable functions such that $f_j \to f$ on T (or even f_j converges almost everywhere on T to f) and $V_\varepsilon(f, T) < \infty$ for all $\varepsilon > 0$. By Egorov's Theorem (e.g., [64, Section 3.2.7]), given $\eta > 0$, there is a measurable set $T_\eta \subset T$ such that $\mathscr{L}(T \setminus T_\eta) \leq \eta$ and $f_j \rightrightarrows f$ on T_η. By (a) above and Lemma 2.2(b), we have

$$\limsup_{j \to \infty} V_\varepsilon(f_j, T_\eta) \leq V_{\varepsilon'}(f, T_\eta) \leq V_{\varepsilon'}(f, T) < \infty \quad \text{for all} \ \ 0 < \varepsilon' < \varepsilon.$$

4.2   Examples illustrating Theorem 4.1

Example 4.7 The main assumption (4.2) in Theorem 4.1 is essential. In fact, it is well known that the sequence of functions $\{f_j\} \subset \mathbb{R}^T$ on the interval $T = [0, 2\pi]$ defined by $f_j(t) = \sin(jt), 0 \leq t \leq 2\pi$, has no subsequence convergent at all points of T (cf. [20, Example 3]; more explicitly this is revived in Remark 4.8 below). Let us show that $\{f_j\}$ does not satisfy condition (4.2).

Let us fix $j \in \mathbb{N}$. First, note that, given $t, s \in [0, 2\pi]$, we have $\sin(jt) = 0$ if and only if $t = t_k = k\pi/j$ for $k = 0, 1, 2, \ldots, 2j$, and $|\sin(js)| = 1$ if and only if $s = s_k = \frac{1}{2}(t_{k-1} + t_k) = (k - \frac{1}{2})\pi/j$ for $k = 1, 2, \ldots, 2j$. Setting $I_k = [t_{k-1}, s_k]$ and $I'_k = [s_k, t_k]$, we find

$$T = [0, 2\pi] = \bigcup_{k=1}^{2j} [t_{k-1}, t_k] = \bigcup_{k=1}^{2j} (I_k \cup I'_k) \quad \text{(non-overlapping intervals)},$$

and f_j is strictly monotone on each interval I_k and I'_k for $k = 1, 2 \ldots, 2j$. By virtue of Lemma 2.8, given $\varepsilon > 0$, we have

$$\sum_{k=1}^{2j} \left(V_\varepsilon(f_j, I_k) + V_\varepsilon(f_j, I'_k) \right) \leq V_\varepsilon(f_j, T) \leq \sum_{k=1}^{2j} \left(V_\varepsilon(f_j, I_k) + V_\varepsilon(f_j, I'_k) \right) + (4j - 1)2\varepsilon.$$

$$(4.12)$$

It suffices to calculate $V_\varepsilon(f_j, I_k)$ for $k = 1$, where $I_1 = [t_0, s_1] = [0, \pi/2j]$ (the other ε-variations in (4.12) are calculated similarly and give the same value). Since f_j is strictly increasing on I_1, $f_j(I_1) = [0, 1]$ and $|f_j(I_1)| = 1$, (3.8) and (3.3) imply $V_\varepsilon(f_j, I_1) = 1 - 2\varepsilon$ if $0 < \varepsilon < \frac{1}{2}$ (and $V_\varepsilon(f_j, I_1) = 0$ if $\varepsilon \geq \frac{1}{2}$). Hence, $V_\varepsilon(f_j, I_k) = V_\varepsilon(f_j, I'_k) = 1 - 2\varepsilon$ for all $0 < \varepsilon < \frac{1}{2}$ and $k = 1, 2, \ldots, 2j$, and it follows from (4.12) that

$$4j(1 - 2\varepsilon) \leq V_\varepsilon(f_j, T) \leq 4j(1 - 2\varepsilon) + (4j - 1)2\varepsilon = 4j - 2\varepsilon, \quad 0 < \varepsilon < \frac{1}{2}.$$

Thus, condition (4.2) is not satisfied by $\{f_j\}$.

Remark 4.8 Since the sequence of functions $\{f_j\}_{j=1}^\infty$ from Example 4.7, i.e., $f_j(t) = \sin(jt)$ for $t \in [0, 2\pi]$, plays a certain role in the sequel as well, for the sake of completeness, we recall here the proof of the fact (e.g., [66,

Chapter 7, Example 7.20]) that no subsequence of $\{\sin(jt)\}_{j=1}^{\infty}$ converges in $\mathbb{R}$ for all $t \in [0, 2\pi]$; note that $\{f_j\}$ is a uniformly bounded sequence of continuous functions on the compact set $[0, 2\pi]$. On the contrary, assume that there is an increasing sequence $\{j_p\}_{p=1}^{\infty} \subset \mathbb{N}$ such that $\sin(j_p t)$ converges as $p \to \infty$ for all $t \in [0, 2\pi]$. Given $t \in [0, 2\pi]$, this implies $\sin(j_p t) - \sin(j_{p+1} t) \to 0$, and so, $(\sin(j_p t) - \sin(j_{p+1} t))^2 \to 0$ as $p \to \infty$. By Lebesgue's dominated convergence theorem,

$$I_p \equiv \int_0^{2\pi} \left(\sin(j_p t) - \sin(j_{p+1} t)\right)^2 dt \to 0 \quad \text{as} \quad p \to \infty.$$

However, a straightforward computation of the integral I_p (note that $j_p < j_{p+1}$) gives the value $I_p = 2\pi$ for all $p \in \mathbb{N}$, which is a contradiction.

More precisely (cf. [66, Chapter 10, Exercise 16]), the set $E \subset [0, 2\pi]$ of all points $t \in [0, 2\pi]$, for which $\sin(j_p t)$ converges as $p \to \infty$ (with $\{j_p\}_{p=1}^{\infty}$ as above), is of Lebesgue measure zero, $\mathcal{L}(E) = 0$. To see this, it suffices to note that, for a measurable set $A \subset E$, $\int_A \sin(j_p t) dt \to 0$ and

$$\int_A (\sin(j_p t))^2 dt = \frac{1}{2} \int_A (1 - \cos(2 j_p t)) dt \to \frac{1}{2} \mathcal{L}(A) \quad \text{as} \quad p \to \infty.$$

To illustrate the assertion in the previous paragraph, let us show that, given $t \in \mathbb{R}$, $\sin(jt)$ converges as $j \to \infty$ if and only if $\sin t = 0$ (i.e., $t = \pi k$ for some integer k). Since the sufficiency is clear, we prove the necessity. Suppose $t \in \mathbb{R}$ and the limit $\phi(t) = \lim_{j \to \infty} \sin(jt)$ exists in $\mathbb{R}$. To show that $\sin t = 0$, we suppose, for contradiction, that $\sin t \neq 0$. Passing to the limit as $j \to \infty$ in the equality

$$\sin(j + 2)t + \sin(jt) = 2 \sin(j + 1)t \cdot \cos t,$$

we get $\phi(t) + \phi(t) = 2\phi(t) \cos t$, which is equivalent to $\phi(t) = 0$ or $\cos t = 1$. Since $\sin(j + 1)t = \sin(jt) \cdot \cos t + \sin t \cdot \cos(jt)$, we find

$$\cos(jt) = \frac{\sin(j + 1)t - \sin(jt) \cdot \cos t}{\sin t},$$

and so, $\lim_{j \to \infty} \cos(jt) = \phi(t)(1 - \cos t)/\sin t$. Hence

$$1 = \lim_{j \to \infty} \left(\sin^2(jt) + \cos^2(jt)\right) = (\phi(t))^2 + (\phi(t))^2 \cdot \left(\frac{1 - \cos t}{\sin t}\right)^2$$

$$= (\phi(t))^2 \cdot \frac{2(1 - \cos t)}{\sin^2 t},$$

and so, $\phi(t) \neq 0$ and $1 \neq \cos t$, which is a contradiction. (In a similar manner, one may show that, given $t \in \mathbb{R}$, $\cos(jt)$ converges as $j \to \infty$ if and only if $\cos t = 1$, i.e., $t = 2\pi k$ for some integer k; see [8, p. 233]).

Returning to the convergence set $E \subset [0, 2\pi]$ of the sequence $\{\sin(j_p t)\}_{p=1}^{\infty}$, as a consequence of the previous assertion, we find that if $j_p = p$, then $E = \{0, \pi, 2\pi\}$. In general, E may be "quite large": for instance, if $j_p = p!$, then $E = \pi \cdot (\mathbb{Q} \cap [0, 2])$, which is countable and dense in $[0, 2\pi]$.

Example 4.9 That condition (4.2) in Theorem 4.1 is *not necessary* for the pointwise convergence $f_j \to f$ can be illustrated by the sequence $\{f_j\}$ from Example 3.9, where $I = [a, b] = [0, 1]$. We assert that

$$\text{if } 0 < \varepsilon < \tfrac{1}{2}d(x, y), \text{ then } \lim_{j \to \infty} V_\varepsilon(f_j, I) = \infty.$$

To see this, given $j \in \mathbb{N}$, we consider a partition of I as defined in (3.41). Supposing that $g \in M^I$ is arbitrary such that $d_{\infty,I}(f_j, g) \leq \varepsilon$, we find, by virtue of (2.2),

$$V(g, I) \geq \sum_{k=1}^{j!} d(g(t_k), g(s_k)) \geq \sum_{k=1}^{j!} (d(f(t_k), f(s_k)) - 2\varepsilon) = j!(d(x, y) - 2\varepsilon).$$

By definition (2.6), $V_\varepsilon(f_j, I) \geq j!(d(x, y) - 2\varepsilon)$, which proves our assertion.

Example 4.10 The choice of an appropriate (equivalent) metric on M is essential in Theorem 4.1. (Recall that two metrics d and d' on M are *equivalent* if, given a sequence $\{x_j\} \subset M$ and $x \in M$, conditions $d(x_j, x) \to 0$ and $d'(x_j, x) \to 0$, as $j \to \infty$, are equivalent.)

Let d be an unbounded metric on M, i.e., $\sup_{x,y \in M} d(x, y) = \infty$ (for instance, given $N \in \mathbb{N}$ and $q \geq 1$, $M = \mathbb{R}^N$, and $d(x, y) = \|x - y\|$, where $x = (x_1, \ldots, x_N)$, $y = (y_1, \ldots, y_N) \in \mathbb{R}^N$ and $\|x\| = (\sum_{i=1}^N |x_i|^q)^{1/q}$). The unboundedness of d is equivalent to $\sup_{x \in M} d(x, y) = \infty$ for all $y \in M$, so let us fix $y_0 \in M$ and pick $\{x_j\} \subset M$ such that $d(x_j, y_0) \to \infty$ as $j \to \infty$ (e.g., in the case $M = \mathbb{R}^N$ we may set $x_j = (j, j, \ldots, j)$ and $y_0 = (0, 0, \ldots, 0)$). Given a sequence $\{\tau_j\} \subset (a, b) \subset I = [a, b]$ such that $\tau_j \to a$ as $j \to \infty$, define $\{f_j\} \subset M^I$ by (cf. Example 3.3)

$$f_j(\tau_j) = x_j \text{ and } f_j(t) = y_0 \text{ if } t \in I \setminus \{\tau_j\}, j \in \mathbb{N}.$$

Clearly, $f_j \to c(t) \equiv y_0$ pointwise on I, and so, $\{f_j\}$ is pointwise relatively compact on I (this can be seen directly by noting that the set $\{f_j(t) : j \in \mathbb{N}\}$ is equal to $\{x_k : k \in \mathbb{N} \text{ and } \tau_k = t\} \cup \{y_0\}$, which is finite for all $t \in I$). Given $\varepsilon > 0$, there is $j_0 = j_0(\varepsilon) \in \mathbb{N}$ such that $|f_j(I)| = d(x_j, y_0) > 2\varepsilon$ for all $j \geq j_0$, and so, by Lemma 2.5(f),

$$V_\varepsilon(f_j, I) \geq |f_j(I)| - 2\varepsilon = d(x_j, y_0) - 2\varepsilon \quad \text{for all} \quad j \geq j_0.$$

Since $\lim_{j \to \infty} d(x_j, y_0) = \infty$, this implies $\lim_{j \to \infty} V_\varepsilon(f_j, I) = \infty$, so Theorem 4.1 is inapplicable in this context.

On the other hand, the metric d' on M, given by $d'(x, y) = \frac{d(x,y)}{1+d(x,y)}$, $x, y \in M$, is equivalent to d. Let us denote by $V'(f_j, I)$ and $V'_\varepsilon(f_j, I)$ the (Jordan) variation and the ε-variation of f_j on I with respect to metric d', respectively. The variation $V'(f_j, I)$ is equal to (by virtue of the additivity of V')

$$V'(f_j, I) = V'(f_j, [a, \tau_j]) + V'(f_j, [\tau_j, b]) = d'(x_j, y_0) + d'(x_j, y_0) = 2 \frac{d(x_j, y_0)}{1 + d(x_j, y_0)}.$$

Now, if $\varepsilon > 0$, by Lemma 2.5(a), $V'_\varepsilon(f_j, I) \leq V'(f_j, I)$ for all $j \in \mathbb{N}$, and so,

$$\limsup_{j \to \infty} V'_\varepsilon(f_j, I) \leq \lim_{j \to \infty} V'(f_j, I) = 2.$$

Thus, the main assumption (4.2) in Theorem 4.1 is satisfied, and this Theorem is applicable to the sequence $\{f_j\}$.

Another interpretation of this example is that the main condition (4.2) is *not invariant* under equivalent metrics on M.

Example 4.11 Here we show that condition (4.2) in Theorem 4.1 is *not necessary* for the pointwise convergence $f_j \to f$ on T, although we have $V_\varepsilon(f_j, T) < \infty$ and $V_\varepsilon(f, T) < \infty$ for all $\varepsilon > 0$. Furthermore, condition (4.2) may not hold with respect to *any* (equivalent) metric on M such that $d(f_j(t), f(t)) \to 0$ as $j \to \infty$ for all $t \in T$. In fact, let $T = [0, 2\pi]$ and $M = \mathbb{R}$, and define $\{f_j\} \subset M^T$ by (cf. [20, Example 4]):

$$f_j(t) = \begin{cases} \sin(j^2 t) & \text{if } 0 \leq t \leq 2\pi/j, \\ 0 & \text{if } 2\pi/j < t \leq 2\pi, \end{cases} \quad j \in \mathbb{N}.$$

Clearly, $\{f_j\}$ converges pointwise on T to the function $f \equiv 0$ with respect to *any* metric d on M, which is equivalent to the usual metric $(x, y) \mapsto |x - y|$ on $\mathbb{R}$. Since f_j is continuous on $T = [0, 2\pi]$ with respect to metric $|x - y|$, and so, with respect to d, we find, by Lemma 2.10, $V_\varepsilon(f_j, T) < \infty$ and $V_\varepsilon(f, T) = 0$ with respect to d for all $j \in \mathbb{N}$ and $\varepsilon > 0$. Now, given $j, n \in \mathbb{N}$, we set

$$s_{j,n} = \frac{1}{j^2}\left(2\pi n - \frac{3\pi}{2}\right) \quad \text{and} \quad t_{j,n} = \frac{1}{j^2}\left(2\pi n - \frac{\pi}{2}\right),$$

so that $f_j(s_{j,n}) = 1$ and $f_j(t_{j,n}) = -1$. Note also that

$$0 < s_{j,1} < t_{j,1} < s_{j,2} < t_{j,2} < \cdots < s_{j,j} < t_{j,j} < 2\pi/j \quad \text{for all} \quad j \in \mathbb{N}.$$

Let $0 < \varepsilon < \frac{1}{2}d(1, -1)$. Given $j \in \mathbb{N}$, suppose $g \in M^T$ is an arbitrary function satisfying condition $d_{\infty,T}(f_j, g) \le \varepsilon$. The definition of $V(g, T)$ and (2.2) give

$$V(g, T) \ge \sum_{n=1}^{j} d(g(s_{j,n}), g(t_{j,n})) \ge \sum_{n=1}^{j} \big(d(f_j(s_{j,n}), f_j(t_{j,n})) - 2\varepsilon \big)$$

$$= \big(d(1, -1) - 2\varepsilon \big) j.$$

By the arbitrariness of g as above and (2.6), $V_\varepsilon(f_j, T) \ge (d(1, -1) - 2\varepsilon) j$, and so, condition (4.2) is not fulfilled for $0 < \varepsilon < \frac{1}{2}d(1, -1)$.

Example 4.12

(a) Theorem 4.1 is inapplicable to the sequence $\{f_j\}$ from Example 3.7, because (although $f_j \rightrightarrows f = \mathscr{D}_{x,y}$ on I) $\lim_{j \to \infty} V_\varepsilon(f_j, I) = \infty$ for $\varepsilon = \frac{1}{2}\|x - y\|$. The reason is that the limit function $\mathscr{D}_{x,y}$ is not regulated (if $x \ne y$). However, see Remark 4.6(a) if the limit function is regulated.

(b) Nevertheless, Theorem 4.1 can be successfully applied to sequences of *nonregulated* functions. To see this, we again use the context of Example 3.7, where we suppose $x = y \in M$, so that $f(t) = \mathscr{D}_{x,x}(t) = x$, $t \in I$. Recall also that we have $f_j = \mathscr{D}_{x_j,y_j}$ with $x_j \ne y_j$, $j \in \mathbb{N}$, $x_j \to x$ and $y_j \to y = x$ in M, and $f_j \rightrightarrows f \equiv x$ on I. Given $\varepsilon > 0$, there is $j_0 = j_0(\varepsilon) \in \mathbb{N}$ such that $\|x_j - y_j\| \le 2\varepsilon$ for all $j \ge j_0$, which implies, by virtue of (3.36), $V_\varepsilon(f_j, I) = 0$ for all $j \ge j_0$. This yields condition (4.2):

$$\limsup_{j \to \infty} V_\varepsilon(f_j, I) \le \sup_{j \ge j_0} V_\varepsilon(f_j, I) = 0$$

(cf. also [29, Example 3]).

On the other hand, for a fixed $k \in \mathbb{N}$ and $0 < \varepsilon < \frac{1}{2}\|x_k - y_k\|$, we have, from (3.36), $V_\varepsilon(f_k, I) = \infty$, and so, $\sup_{j \in \mathbb{N}} V_\varepsilon(f_j, I) \ge V_\varepsilon(f_k, I) = \infty$. Thus, condition of uniform boundedness of ε-variations $\sup_{j \in \mathbb{N}} V_\varepsilon(f_j, I) < \infty$, which was assumed in [43, Theorem 3.8], is more restrictive than condition (4.2).

4.3   Two extensions of Theorem 4.1

Applying Theorem 4.1 and the diagonal procedure over expanding intervals, we get the following *local* version of Theorem 4.1.

Theorem 4.13 *If $T \subset \mathbb{R}$, (M, d) is a metric space, and $\{f_j\} \subset M^T$ is a pointwise relatively compact sequence of functions such that*

$$\limsup_{j \to \infty} V_\varepsilon(f_j, T \cap [a, b]) < \infty \text{ for all } a, b \in T, \ a \le b, \text{ and } \varepsilon > 0,$$

then a subsequence of $\{f_j\}$ *converges pointwise on* T *to a regulated function* $f \in$ *Reg*$(T; M)$ *such that* f *is bounded on* $T \cap [a, b]$ *for all* $a, b \in T, a \leq b$.

Proof With no loss of generality, we may assume that sequences $\{a_k\}$ and $\{b_k\}$ from T are such that $a_{k+1} < a_k < b_k < b_{k+1}$ for all $k \in \mathbb{N}$, $a_k \to \inf T \notin T$ and $b_k \to \sup T \notin T$ as $k \to \infty$. By Theorem 4.1, applied to $\{f_j\}$ on $T \cap [a_1, b_1]$, there is a subsequence $\{J_1(j)\}_{j=1}^{\infty}$ of $\{J_0(j)\}_{j=1}^{\infty} = \{j\}_{j=1}^{\infty}$ such that $\{f_{J_1(j)}\}_{j=1}^{\infty}$ converges pointwise on $T \cap [a_1, b_1]$ to a bounded regulated function $f_1' : T \cap [a_1, b_1] \to M$. Since

$$\limsup_{j \to \infty} V_\varepsilon(f_{J_1(j)}, T \cap [a_2, b_2]) \leq \limsup_{j \to \infty} V_\varepsilon(f_j, T \cap [a_2, b_2]) < \infty \quad \text{for all} \ \varepsilon > 0,$$

applying Theorem 4.1 to $\{f_{J_1(j)}\}_{j=1}^{\infty}$ on $T \cap [a_2, b_2]$, we find a subsequence $\{J_2(j)\}_{j=1}^{\infty}$ of $\{J_1(j)\}_{j=1}^{\infty}$ such that $\{f_{J_2(j)}\}_{j=1}^{\infty}$ converges pointwise on the set $T \cap [a_2, b_2]$ to a bounded regulated function $f_2' : T \cap [a_2, b_2] \to M$. Since $[a_1, b_1] \subset [a_2, b_2]$, we get $f_2'(t) = f_1'(t)$ for all $t \in T \cap [a_1, b_1]$. Proceeding this way, for each $k \in \mathbb{N}$ we obtain a subsequence $\{J_k(j)\}_{j=1}^{\infty}$ of $\{J_{k-1}(j)\}_{j=1}^{\infty}$ and a bounded regulated function $f_k' : T \cap [a_k, b_k] \to M$ such that $f_{J_k(j)} \to f_k'$ on $T \cap [a_k, b_k]$ as $j \to \infty$ and $f_k'(t) = f_{k-1}'(t)$ for all $t \in T \cap [a_{k-1}, b_{k-1}]$. Define $f : T \to M$ as follows: given $t \in T$, we have $\inf T < t < \sup T$, so there is $k = k(t) \in \mathbb{N}$ such that $t \in T \cap [a_k, b_k]$, and so, we set $f(t) = f_k'(t)$. The diagonal sequence $\{f_{J_j(j)}\}_{j=1}^{\infty}$ converges pointwise on T to the function f, which satisfies the *conclusions* of Theorem 4.13. $\qquad \square$

Theorem 4.1 implies immediately that if (M, d) is a *proper* metric space, $\{f_j\} \subset M^T$ is a pointwise relatively compact sequence, and there is $E \subset T$ of measure zero, $\mathscr{L}(E) = 0$, such that $\limsup_{j \to \infty} V_\varepsilon(f_j, T \setminus E) < \infty$ for all $\varepsilon > 0$, then a subsequence of $\{f_j\}$ converges a.e. (= almost everywhere) on T to a function $f \in M^T$ such that $V_\varepsilon(f, T \setminus E) < \infty$ for all $\varepsilon > 0$. The following theorem is a *selection principle for the a.e. convergence* (it may be considered as subsequence-converse to Remark 4.6(b) concerning the "almost necessity" of condition (4.2) for the pointwise convergence).

Theorem 4.14 *Suppose* $T \subset \mathbb{R}$, (M, d) *is a proper metric space, and* $\{f_j\} \subset M^T$ *is a pointwise relatively compact (or a.e. relatively compact) on* T *sequence of functions satisfying the condition: for every* $\eta > 0$ *there is a measurable set* $E_\eta \subset T$ *of Lebesgue measure* $\mathscr{L}(E_\eta) \leq \eta$ *such that*

$$\limsup_{j \to \infty} V_\varepsilon(f_j, T \setminus E_\eta) < \infty \quad \text{for all} \ \ \varepsilon > 0. \tag{4.13}$$

Then, a subsequence of $\{f_j\}$ *converges a.e. on* T *to a function* $f \in M^T$ *having the property: given* $\eta > 0$, *there is a measurable set* $E_\eta' \subset T$ *of Lebesgue measure* $\mathscr{L}(E_\eta') \leq \eta$ *such that* $V_\varepsilon(f, T \setminus E_\eta') < \infty$ *for all* $\varepsilon > 0$.

Proof We follow the proof of Theorem 6 from [20] with appropriate modifications. Let $T_0 \subset T$ be a set of Lebesgue measure zero such that the set $\{f_j(t) : j \in \mathbb{N}\}$ is relatively compact in M for all $t \in T \setminus T_0$. We employ Theorem 4.1 several times as well as the diagonal procedure. By the assumption, there is a measurable set $E_1 \subset T$ of measure $\mathscr{L}(E_1) \leq 1$ such that (4.13) holds with $\eta = 1$. The sequence $\{f_j\}$ is pointwise relatively compact on $T \setminus (T_0 \cup E_1)$ and, by Lemma 2.2(b), for all $\varepsilon > 0$,

$$\limsup_{j\to\infty} V_\varepsilon(f_j, T \setminus (T_0 \cup E_1)) \leq \limsup_{j\to\infty} V_\varepsilon(f_j, T \setminus E_1) < \infty.$$

By Theorem 4.1, there are a subsequence $\{J_1(j)\}_{j=1}^\infty$ of $\{j\}_{j=1}^\infty$ and a function $f^{(1)} : T \setminus (T_0 \cup E_1) \to M$, satisfying $V_\varepsilon(f^{(1)}, T \setminus (T_0 \cup E_1)) < \infty$ for all $\varepsilon > 0$, such that $f_{J_1(j)} \to f^{(1)}$ pointwise on $T \setminus (T_0 \cup E_1)$ as $j \to \infty$. Inductively, if $k \geq 2$ and a subsequence $\{J_{k-1}(j)\}_{j=1}^\infty$ of $\{j\}_{j=1}^\infty$ is already chosen, by the assumption (4.13), there is a measurable set $E_k \subset T$ with $\mathscr{L}(E_k) \leq 1/k$ such that $\limsup_{j\to\infty} V_\varepsilon(f_j, T \setminus E_k)$ is finite for all $\varepsilon > 0$. The sequence $\{f_{J_{k-1}(j)}\}_{j=1}^\infty$ is pointwise relatively compact on $T \setminus (T_0 \cup E_k)$ and, again by Lemma 2.2(b), for all $\varepsilon > 0$,

$$\limsup_{j\to\infty} V_\varepsilon(f_{J_{k-1}(j)}, T \setminus (T_0 \cup E_k)) \leq \limsup_{j\to\infty} V_\varepsilon(f_{J_{k-1}(j)}, T \setminus E_k)$$

$$\leq \limsup_{j\to\infty} V_\varepsilon(f_j, T \setminus E_k) < \infty.$$

Theorem 4.1 implies the existence of a subsequence $\{J_k(j)\}_{j=1}^\infty$ of $\{J_{k-1}(j)\}_{j=1}^\infty$ and a function $f^{(k)} : T \setminus (T_0 \cup E_k) \to M$, satisfying $V_\varepsilon(f^{(k)}, T \setminus (T_0 \cup E_k)) < \infty$ for all $\varepsilon > 0$, such that $f_{J_k(j)} \to f^{(k)}$ pointwise on $T \setminus (T_0 \cup E_k)$ as $j \to \infty$.

The set $E = T_0 \cup \bigcap_{k=1}^\infty E_k$ is of measure zero, and we have the equality $T \setminus E = \bigcup_{k=1}^\infty (T \setminus (T_0 \cup E_k))$. Define the function $f : T \setminus E \to M$ as follows: given $t \in T \setminus E$, there is $k \in \mathbb{N}$ such that $t \in T \setminus (T_0 \cup E_k)$, so we set $f(t) = f^{(k)}(t)$. The value $f(t)$ is well-defined, i.e., it is independent of a particular k: in fact, if $t \in T \setminus (T_0 \cup E_{k_1})$ for some $k_1 \in \mathbb{N}$ with, say, $k < k_1$ (with no loss of generality), then, by the construction above, $\{J_{k_1}(j)\}_{j=1}^\infty$ is a subsequence of $\{J_k(j)\}_{j=1}^\infty$, which implies

$$f^{(k_1)}(t) = \lim_{j\to\infty} f_{J_{k_1}(j)}(t) = \lim_{j\to\infty} f_{J_k(j)}(t) = f^{(k)}(t) \quad \text{in} \quad M.$$

Let us show that the diagonal sequence $\{f_{J_j(j)}\}_{j=1}^\infty$ (which, of course, is a subsequence of the original sequence $\{f_j\}$) converges pointwise on $T \setminus E$ to the function f. To see this, suppose $t \in T \setminus E$. Then $t \in T \setminus (T_0 \cup E_k)$ for some $k \in \mathbb{N}$, and so, $f(t) = f^{(k)}(t)$. Since $\{f_{J_j(j)}\}_{j=k}^\infty$ is a subsequence of $\{f_{J_k(j)}\}_{j=1}^\infty$, we find

$$\lim_{j\to\infty} f_{J_j(j)}(t) = \lim_{j\to\infty} f_{J_k(j)}(t) = f^{(k)}(t) = f(t) \quad \text{in} \quad M.$$

We extend f from $T \setminus E$ to the whole T arbitrarily and denote this extension again by f. Given $\eta > 0$, pick the smallest $k \in \mathbb{N}$ such that $1/k \leq \eta$ and set $E'_\eta = T_0 \cup E_k$. It follows that $\mathscr{L}(E'_\eta) = \mathscr{L}(E_k) \leq 1/k \leq \eta$, $f = f^{(k)}$ on $T \setminus (T_0 \cup E_k) = T \setminus E'_\eta$, and

$$V_\varepsilon(f, T \setminus E'_\eta) = V_\varepsilon(f^{(k)}, T \setminus (T_0 \cup E_k)) < \infty \quad \text{for all} \quad \varepsilon > 0,$$

which was to be proved. $\qquad\qquad\qquad\qquad\qquad\qquad\qquad\qquad\qquad\qquad\qquad\square$

4.4 Weak pointwise selection principles

In this section, we establish a variant of Theorem 4.1 for functions with values in a reflexive Banach space taking into account some specific features of this case (such as the validity of the weak pointwise convergence of sequences of functions).

Suppose $(M, \|\cdot\|)$ is a normed linear space over the field $\mathbb{K} = \mathbb{R}$ or $\mathbb{C}$ (equipped with the absolute value $|\cdot|$) and M^* is its *dual*, i.e., $M^* = L(M; \mathbb{K})$ is the space of all continuous (= bounded) linear functionals on M. Recall that M^* is a Banach space under the norm

$$\|x^*\| = \sup\{|x^*(x)| : x \in M \text{ and } \|x\| \leq 1\}, \quad x^* \in M^*. \qquad (4.14)$$

The natural duality between M and M^* is determined by the bilinear functional $\langle \cdot, \cdot \rangle : M \times M^* \to \mathbb{K}$ defined by $\langle x, x^* \rangle = x^*(x)$ for all $x \in M$ and $x^* \in M^*$. Recall also that a sequence $\{x_j\} \subset M$ is said to *converge weakly* in M to an element $x \in M$, written as $x_j \overset{w}{\to} x$ in M, if $\langle x_j, x^* \rangle \to \langle x, x^* \rangle$ in $\mathbb{K}$ as $j \to \infty$ for all $x^* \in M^*$. It is well known that if $x_j \overset{w}{\to} x$ in M, then $\|x\| \leq \liminf_{j \to \infty} \|x_j\|$.

The notion of the approximate variation $\{V_\varepsilon(f, T)\}_{\varepsilon > 0}$ for $f \in M^T$ is introduced as in (2.6) with respect to the induced metrics $d(x, y) = \|x - y\|$, $x, y \in M$, and $d_{\infty,T}(f, g) = \|f - g\|_{\infty,T} = \sup_{t \in T} \|f(t) - g(t)\|$, $f, g \in M^T$.

Theorem 4.15 *Let $T \subset \mathbb{R}$ and $(M, \|\cdot\|)$ be a reflexive Banach space with separable dual $(M^*, \|\cdot\|)$. Suppose the sequence of functions $\{f_j\} \subset M^T$ is such that*

(i) $\sup_{j \in \mathbb{N}} \|f_j(t_0)\| \leq C_0$ *for some $t_0 \in T$ and $C_0 \geq 0$;*
(ii) $v(\varepsilon) \equiv \limsup_{j \to \infty} V_\varepsilon(f_j, T) < \infty$ *for all $\varepsilon > 0$.*

Then, there is a subsequence of $\{f_j\}$, again denoted by $\{f_j\}$, and a function $f \in M^T$, satisfying $V_\varepsilon(f, T) \leq v(\varepsilon)$ for all $\varepsilon > 0$ (and, a fortiori, f is bounded and regulated on T), such that $f_j(t) \overset{w}{\to} f(t)$ in M for all $t \in T$.

Proof

1. First, we show that there is $j_0 \in \mathbb{N}$ such that $C(t) \equiv \sup_{j \geq j_0} \|f_j(t)\|$ is finite for all $t \in T$. In fact, by Lemma 2.5(b), $|f_j(T)| \leq V_\varepsilon(f_j, T) + 2\varepsilon$ with, say, $\varepsilon = 1$, for all $j \in \mathbb{N}$, which implies

$$\limsup_{j \to \infty} |f_j(T)| \le \limsup_{j \to \infty} V_1(f_j, T) + 2 = v(1) + 2 < \infty \quad \text{by (ii)},$$

and so, there is $j_0 \in \mathbb{N}$ and a constant $C_1 > 0$ such that $|f_j(T)| \le C_1$ for all $j \ge j_0$. Now, given $j \ge j_0$ and $t \in T$, we get, by (i),

$$\|f_j(t)\| \le \|f_j(t_0)\| + \|f_j(t) - f_j(t_0)\| \le C_0 + |f_j(T)| \le C_0 + C_1,$$

i.e., $C(t) \le C_0 + C_1$ for all $t \in T$.

2. Given $j \in \mathbb{N}$ and $x^* \in M^*$, we set $f_j^{x^*}(t) = \langle f_j(t), x^* \rangle$ for all $t \in T$. Let us verify that the sequence $\{f_j^{x^*}\}_{j=j_0}^{\infty} \subset \mathbb{K}^T$ satisfies the assumptions of Theorem 4.1. By (4.14) and Step 1, we have

$$|f_j^{x^*}(t)| \le \|f_j(t)\| \cdot \|x^*\| \le C(t)\|x^*\| \quad \text{for all } t \in T \text{ and } j \ge j_0, \qquad (4.15)$$

and so, $\{f_j^{x^*}\}_{j=j_0}^{\infty}$ is pointwise relatively compact on T. If $x^* = 0$, then $f_j^{x^*} = 0$ in $\mathbb{K}^T$, which implies $V_\varepsilon(f_j^{x^*}, T) = 0$ for all $j \in \mathbb{N}$ and $\varepsilon > 0$. Now, we show that if $x^* \ne 0$, then

$$V_\varepsilon(f_j^{x^*}, T) \le V_{\varepsilon/\|x^*\|}(f_j, T)\|x^*\| \quad \text{for all } j \in \mathbb{N} \text{ and } \varepsilon > 0. \qquad (4.16)$$

To prove (4.16), we may assume that $V_{\varepsilon/\|x^*\|}(f_j, T) < \infty$. By definition (2.6), for every $\eta > 0$ there is $g_j = g_{j,\eta} \in BV(T; M)$ (also depending on ε and x^*) such that

$$\|f_j - g_j\|_{\infty,T} \le \varepsilon/\|x^*\| \quad \text{and} \quad V(g_j, T) \le V_{\varepsilon/\|x^*\|}(f_j, T) + \eta.$$

Setting $g_j^{x^*}(t) = \langle g_j(t), x^* \rangle$ for all $t \in T$ (and so, $g_j^{x^*} \in \mathbb{K}^T$), we find

$$\left| f_j^{x^*} - g_j^{x^*} \right|_{\infty,T} = \sup_{t \in T} |\langle f_j(t) - g_j(t), x^* \rangle| \le \sup_{t \in T} \|f_j(t) - g_j(t)\| \cdot \|x^*\|$$

$$= \|f_j - g_j\|_{\infty,T}\|x^*\| \le (\varepsilon/\|x^*\|)\|x^*\| = \varepsilon.$$

Furthermore, it is straightforward that $V(g_j^{x^*}, T) \le V(g_j, T)\|x^*\|$. Once again from definition (2.6), it follows that

$$V_\varepsilon(f_j^{x^*}, T) \le V(g_j^{x^*}, T) \le V(g_j, T)\|x^*\| \le \left(V_{\varepsilon/\|x^*\|}(f_j, T) + \eta\right)\|x^*\|.$$

Passing to the limit as $\eta \to +0$, we arrive at (4.16). Now, by (4.16) and (ii), for every $\varepsilon > 0$ and $x^* \in M^*$, $x^* \neq 0$, we have

$$v_{x^*}(\varepsilon) \equiv \limsup_{j \to \infty} V_\varepsilon(f_j^{x^*}, T) \leq v(\varepsilon/\|x^*\|)\|x^*\| < \infty. \tag{4.17}$$

Taking into account (4.15) and (4.17), given $x^* \in M^*$, we may apply Theorem 4.1 to the sequence $\{f_j^{x^*}\}_{j=j_0}^\infty \subset \mathbb{K}^T$ and extract a subsequence $\{f_{j,x^*}\}_{j=1}^\infty$ (depending on x^* as well) of $\{f_j\}_{j=j_0}^\infty$ and find a function $f_{x^*} \in \mathbb{K}^T$, satisfying $V_\varepsilon(f_{x^*}, T) \leq v_{x^*}(\varepsilon)$ for all $\varepsilon > 0$ (and so, f_{x^*} is bounded and regulated on T), such that $\langle f_{j,x^*}(t), x^* \rangle \to f_{x^*}(t)$ in $\mathbb{K}$ as $j \to \infty$ for all $t \in T$.

3. Making use of the diagonal procedure, we are going to get rid of the dependence of $\{f_{j,x^*}\}_{j=1}^\infty$ on the element $x^* \in M^*$. Since M^* is separable, M^* contains a countable dense subset $\{x_k^*\}_{k=1}^\infty$. Setting $x^* = x_1^*$ in (4.15) and (4.17) and applying Theorem 4.1 to the sequence of functions $f_j^{x_1^*} = \langle f_j(\cdot), x_1^* \rangle \in \mathbb{K}^T$, $j \geq j_0$, we obtain a subsequence $\{J_1(j)\}_{j=1}^\infty$ of $\{j\}_{j=j_0}^\infty$ and a function $f_{x_1^*} \in \mathbb{K}^T$ (both depending on x_1^*), satisfying $V_\varepsilon(f_{x_1^*}, T) \leq v_{x_1^*}(\varepsilon)$ for all $\varepsilon > 0$, such that $\langle f_{J_1(j)}(t), x_1^* \rangle \to f_{x_1^*}(t)$ in $\mathbb{K}$ as $j \to \infty$ for all $t \in T$. Inductively, assume that $k \geq 2$ and a subsequence $\{J_{k-1}(j)\}_{j=1}^\infty$ of $\{j\}_{j=j_0}^\infty$ is already chosen. Putting $x^* = x_k^*$, replacing j by $J_{k-1}(j)$ in (4.15), and taking into account (4.17), we get

$$\sup_{j \in \mathbb{N}} |\langle f_{J_{k-1}(j)}(t), x_k^* \rangle| \leq C(t)\|x_k^*\| \quad \text{for all} \quad t \in T$$

and

$$\limsup_{j \to \infty} V_\varepsilon(\langle f_{J_{k-1}(j)}(\cdot), x_k^* \rangle, T) \leq v_{x_k^*}(\varepsilon) < \infty \quad \text{for all} \quad \varepsilon > 0.$$

By Theorem 4.1, applied to the sequence $\{f_{J_{k-1}(j)}^{x_k^*}\}_{j=1}^\infty \subset \mathbb{K}^T$, there are a subsequence $\{J_k(j)\}_{j=1}^\infty$ of $\{J_{k-1}(j)\}_{j=1}^\infty$ and a function $f_{x_k^*} \in \mathbb{K}^T$, satisfying $V_\varepsilon(f_{x_k^*}, T) \leq v_{x_k^*}(\varepsilon)$ for all $\varepsilon > 0$, such that $\langle f_{J_k(j)}(t), x_k^* \rangle \to f_{x_k^*}(t)$ in $\mathbb{K}$ as $j \to \infty$ for all $t \in T$. It follows that the diagonal subsequence $\{f_{J_j(j)}\}_{j=1}^\infty$ of $\{f_j\}_{j=j_0}^\infty$, denoted by $\{f_j\}_{j=1}^\infty$, satisfies the condition:

$$\lim_{j \to \infty} \langle f_j(t), x_k^* \rangle = f_{x_k^*}(t) \quad \text{for all} \quad t \in T \text{ and } k \in \mathbb{N}. \tag{4.18}$$

4. Let us show that the sequence $f_j^{x^*}(t) = \langle f_j(t), x^* \rangle$, $j \in \mathbb{N}$, is Cauchy in $\mathbb{K}$ for every $x^* \in M^*$ and $t \in T$. Since the sequence $\{x_k^*\}_{k=1}^\infty$ is dense in M^*, given $\eta > 0$, there is $k = k(\eta) \in \mathbb{N}$ such that $\|x^* - x_k^*\| \leq \eta/(4C(t) + 1)$, and, by (4.18), there is $j^0 = j^0(\eta) \in \mathbb{N}$ such that $|\langle f_j(t), x_k^* \rangle - \langle f_{j'}(t), x_k^* \rangle| \leq \eta/2$ for all $j, j' \geq j^0$. Hence

$$|f_j^{x^*}(t) - f_{j'}^{x^*}(t)| \leq \|f_j(t) - f_{j'}(t)\| \cdot \|x^* - x_k^*\| + |\langle f_j(t), x_k^* \rangle - \langle f_{j'}(t), x_k^* \rangle|$$

$$\leq 2C(t) \frac{\eta}{4C(t) + 1} + \frac{\eta}{2} \leq \eta \quad \text{for all} \quad j, j' \geq j^0.$$

By the completeness of $(\mathbb{K}, |\cdot|)$, there is $f_{x^*}(t) \in \mathbb{K}$ such that $f_j^{x^*}(t) \to f_{x^*}(t)$ in $\mathbb{K}$ as $j \to \infty$. Thus, we have shown that for every $x^* \in M^*$ there is a function $f_{x^*} \in \mathbb{K}^T$ satisfying, by virtue of Lemma 2.14(c), (4.16) and (4.17),

$$V_\varepsilon(f_{x^*}, T) \leq \liminf_{j \to \infty} V_\varepsilon(f_j^{x^*}, T) \leq v_{x^*}(\varepsilon) \quad \text{for all} \quad \varepsilon > 0$$

and such that

$$\lim_{j \to \infty} \langle f_j(t), x^* \rangle = f_{x^*}(t) \text{ in } \mathbb{K} \text{ for all } t \in T. \tag{4.19}$$

5. Now, we show that, for every $t \in T$, the sequence $\{f_j(t)\}$ converges weakly in M to an element of M. The reflexivity of M implies

$$f_j(t) \in M = M^{**} = L(M^*; \mathbb{K}) \quad \text{for all} \quad j \in \mathbb{N}.$$

Define the functional $F_t : M^* \to \mathbb{K}$ by $F_t(x^*) = f_{x^*}(t)$ for all $x^* \in M^*$. It follows from (4.19) that

$$\lim_{j \to \infty} \langle f_j(t), x^* \rangle = f_{x^*}(t) = F_t(x^*) \quad \text{for all} \quad x^* \in M^*,$$

i.e., the sequence of functionals $\{f_j(t)\} \subset L(M^*; \mathbb{K})$ converges pointwise on M^* to the functional $F_t : M^* \to \mathbb{K}$. By the uniform boundedness principle, $F_t \in L(M^*; \mathbb{K})$ and $\|F_t\| \leq \liminf_{j \to \infty} \|f_j(t)\|$. Setting $f(t) = F_t$ for all $t \in T$, we find $f \in M^T$ and, for all $x^* \in M^*$ and $t \in T$,

$$\lim_{j \to \infty} \langle f_j(t), x^* \rangle = F_t(x^*) = \langle F_t, x^* \rangle = \langle f(t), x^* \rangle, \tag{4.20}$$

which means that $f_j(t) \xrightarrow{w} f(t)$ in M for all $t \in T$. (Note that (4.19) and (4.20) imply $f_{x^*}(t) = \langle f(t), x^* \rangle$ for all $x^* \in M^*$ and $t \in T$.)

6. It remains to prove that $V_\varepsilon(f, T) \leq v(\varepsilon)$ for all $\varepsilon > 0$. Recall that the sequence $\{f_j\} \subset M^T$, we deal with here, is the diagonal sequence $\{f_{J_j(j)}\}_{j=1}^\infty$ from the end of Step 3, which satisfies conditions (4.20) and, in place of (ii),

$$\limsup_{j \to \infty} V_\varepsilon(f_j, T) \leq v(\varepsilon) \quad \text{for all} \quad \varepsilon > 0. \tag{4.21}$$

Let us fix $\varepsilon > 0$. Since $v(\varepsilon) < \infty$ by (ii), for every $\eta > v(\varepsilon)$ condition (4.21) implies the existence of $j_1 = j_1(\eta, \varepsilon) \in \mathbb{N}$ such that $\eta > V_\varepsilon(f_j, T)$ for all $j \geq j_1$. Hence, for every $j \geq j_1$, by the definition of $V_\varepsilon(f_j, T)$, there is $g_j \in BV(T; M)$ such that

$$\|f_j - g_j\|_{\infty, T} = \sup_{t \in T} \|f_j(t) - g_j(t)\| \leq \varepsilon \quad \text{and} \quad V(g_j, T) \leq \eta. \qquad (4.22)$$

These conditions and assumption (i) imply $\sup_{j \geq j_1} V(g_j, T) \leq \eta$ and

$$\|g_j(t_0)\| \leq \|g_j(t_0) - f_j(t_0)\| + \|f_j(t_0)\| \leq \|g_j - f_j\|_{\infty, T} + C_0 \leq \varepsilon + C_0$$

for all $j \geq j_1$. Since $(M, \|\cdot\|)$ is a reflexive Banach space with separable dual M^*, by the weak Helly-type pointwise selection principle (see Theorem 7 and Remarks (1)–(4) in [20], or Theorem 3.5 in [5, Chapter 1]), there are a subsequence $\{g_{j_p}\}_{p=1}^{\infty}$ of $\{g_j\}_{j=j_1}^{\infty}$ and a function $g \in BV(T; M)$ such that $g_{j_p} \xrightarrow{w} g(t)$ in M as $p \to \infty$ for all $t \in T$. Noting that $f_{j_p}(t) \xrightarrow{w} f(t)$ in M as $p \to \infty$ for all $t \in T$ as well, we get $f_{j_p}(t) - g_{j_p}(t) \xrightarrow{w} f(t) - g(t)$ in M as $p \to \infty$, and so, taking into account the first condition in (4.22), we find

$$\|f(t) - g(t)\| \leq \liminf_{p \to \infty} \|f_{j_p}(t) - g_{j_p}(t)\| \leq \varepsilon \quad \text{for all} \quad t \in T,$$

which implies $\|f - g\|_{\infty, T} \leq \varepsilon$. Had we already shown that $V(g, T) \leq \eta$, definition (2.6) would yield $V_\varepsilon(f, T) \leq V(g, T) \leq \eta$ for every $\eta > v(\varepsilon)$, which completes the proof of Theorem 4.15.

In order to prove that $V(g, T) \leq \eta$, suppose $P = \{t_i\}_{i=0}^{m} \subset T$ is a partition of T. Since $g_{j_p}(t) \xrightarrow{w} g(t)$ in M as $p \to \infty$ for all $t \in T$, given $i \in \{1, 2, \ldots, m\}$, we have $g_{j_p}(t_i) - g_{j_p}(t_{i-1}) \xrightarrow{w} g(t_i) - g(t_{i-1})$ in M as $p \to \infty$, and so,

$$\|g(t_i) - g(t_{i-1})\| \leq \liminf_{p \to \infty} \|g_{j_p}(t_i) - g_{j_p}(t_{i-1})\|.$$

Summing over $i = 1, 2, \ldots, m$ and taking into account the properties of the limit inferior and the second condition in (4.22), we get

$$\sum_{i=1}^{m} \|g(t_i) - g(t_{i-1})\| \leq \sum_{i=1}^{m} \liminf_{p \to \infty} \|g_{j_p}(t_i) - g_{j_p}(t_{i-1})\|$$

$$\leq \liminf_{p \to \infty} \sum_{i=1}^{m} \|g_{j_p}(t_i) - g_{j_p}(t_{i-1})\|$$

$$\leq \liminf_{p \to \infty} V(g_{j_p}, T) \leq \eta.$$

Thus, by the arbitrariness of partition P of T, we conclude that $V(g, T) \le \eta$, which was to be proved.   □

Assumption (ii) in Theorem 4.15 can be weakened as the following theorem shows.

Theorem 4.16 *Under the assumptions of Theorem 4.15 on T and $(M, \| \cdot \|)$, suppose the sequence $\{f_j\} \subset M^T$ is such that*

(i) $C(t) \equiv \sup_{j \in \mathbb{N}} \|f_j(t)\| < \infty$ *for all $t \in T$;*
(ii) $v_{x^*}(\varepsilon) \equiv \limsup_{j \to \infty} V_\varepsilon(\langle f_j(\cdot), x^* \rangle, T) < \infty$ *for all $\varepsilon > 0$ and $x^* \in M^*$.*[2]

Then, there is a subsequence of $\{f_j\}$, again denoted by $\{f_j\}$, and a function $f \in M^T$, satisfying $V_\varepsilon(\langle f(\cdot), x^* \rangle, T) \le v_{x^*}(\varepsilon)$ for all $\varepsilon > 0$ and $x^* \in M^*$, such that $f_j(t) \overset{w}{\to} f(t)$ in M as $j \to \infty$ for all $t \in T$.

Proof It suffices to note that assumption (i) implies (4.15) with $j_0 = 1$, replace (4.17) by assumption (ii), and argue as in Steps 3–5 of the proof of Theorem 4.15.
                                                                                    □

The next example illustrates the applicability of Theorems 4.15 and 4.16.

Example 4.17 In examples (a) and (b) below, we assume the following. Let $M = L^2[0, 2\pi]$ be the real Hilbert space of all square Lebesgue summable functions on the closed interval $[0, 2\pi]$ equipped with the *inner product*

$$\langle x, y \rangle = \int_0^{2\pi} x(s)y(s)\, ds \quad \text{and the } norm \quad \|x\| = \sqrt{\langle x, x \rangle}, \quad x, y \in M.$$

It is well known that M is separable, self-adjoint $(M = M^*)$, and so, reflexive $(M = M^{**})$. Given $j \in \mathbb{N}$, define two functions $x_j, y_j \in M$ by

$$x_j(s) = \sin(js) \quad \text{and} \quad y_j(s) = \cos(js) \quad \text{for all} \quad s \in [0, 2\pi].$$

Clearly, $\|x_j\| = \|y_j\| = \sqrt{\pi}$ and, by Lyapunov–Parseval's equality,

$$\frac{\langle x, 1 \rangle^2}{8} + \sum_{j=1}^{\infty} \left(\langle x, x_j \rangle^2 + \langle x, y_j \rangle^2 \right) = \pi \|x\|^2, \quad x \in M,$$

we find $\langle x, x_j \rangle \to 0$ and $\langle x, y_j \rangle \to 0$ as $j \to \infty$ for all $x \in M$, and so, $x_j \overset{w}{\to} 0$ and $y_j \overset{w}{\to} 0$ in M.

In examples (a) and (b) below, we set $T = I = [0, 1]$.

(a) This example illustrates Theorem 4.15. Define the sequence $\{f_j\} \subset M^T$ by $f_j(t) = t x_j$, $t \in T$. Clearly, $f_j(t) \overset{w}{\to} 0$ in M for all $t \in T$. Note, however, that

---

[2] As in Step 2 of the proof of Theorem 4.15, $\langle f_j(\cdot), x^* \rangle(t) = \langle f_j(t), x^* \rangle = f_j^{x^*}(t), t \in T$.

the sequence $\{f_j(t)\}$ does *not* converge in (the norm of) M at all points $0 < t \le 1$, because $\|f_j(t) - f_k(t)\|^2 = (\|x_j\|^2 + \|x_k\|^2)t^2 = 2\pi t^2$, $j \ne k$.

Since $f_j(0) = 0$ in M for all $j \in \mathbb{N}$, we verify only condition (ii) of Theorem 4.15. Setting $\varphi(t) = t$ for $t \in T$, $x = x_j$, and $y = 0$ in (3.1), we find $|\varphi(T)| = 1$ and $\|x\| = \|x_j\| = \sqrt{\pi}$, and so, by virtue of (3.8), we get

$$V_\varepsilon(f_j, T) = \begin{cases} \sqrt{\pi} - 2\varepsilon & \text{if } 0 < \varepsilon < \sqrt{\pi}/2 \\ 0 & \text{if } \varepsilon \ge \sqrt{\pi}/2 \end{cases} \quad \text{for all} \quad j \in \mathbb{N},$$

which implies condition (ii) in Theorem 4.15. Note that (cf. Lemma 2.5(a))

$$V(f_j, T) = \lim_{\varepsilon \to +0} V_\varepsilon(f_j, T) = \sqrt{\pi} \quad \text{for all} \quad j \in \mathbb{N}.$$

Also, it is to be noted that Theorem 4.1 is inapplicable to $\{f_j\}$, because the set $\{f_j(t) : j \in \mathbb{N}\}$ is not relatively compact in (the norm of) M for all $0 < t \le 1$.

(b) Here we present an example when Theorem 4.16 is applicable, while Theorem 4.15 is not. Taking into account definition (3.28) of the Dirichlet function, we let the sequence $\{f_j\} \subset M^T$ be given by $f_j(t) = \mathscr{D}_{x_j, y_j}(t)$ for all $t \in T$ and $j \in \mathbb{N}$. More explicitly,

$$f_j(t)(s) = \mathscr{D}_{x_j(s), y_j(s)}(t) = \begin{cases} \sin(js) & \text{if } t \in I_1 \equiv [0, 1] \cap \mathbb{Q}, \\ \cos(js) & \text{if } t \in I_2 \equiv [0, 1] \setminus \mathbb{Q}, \end{cases} \quad s \in [0, 2\pi].$$

Note that

$$f_j(t) = \mathscr{D}_{x_j, 0}(t) + \mathscr{D}_{0, y_j}(t) = \mathscr{D}_{1,0}(t)x_j + \mathscr{D}_{0,1}(t)y_j, \quad t \in T, \tag{4.23}$$

where $\mathscr{D}_{1,0}$ and $\mathscr{D}_{0,1}$ are the corresponding real-valued Dirichlet functions on the interval $T = [0, 1]$. By (4.23), $f_j(t) \xrightarrow{w} 0$ in M for all $t \in T$. On the other hand, the sequence $\{f_j(t)\}$ *diverges* in (the norm of) M at all points $t \in T$: in fact,

$$\|x_j - x_k\|^2 = \langle x_j - x_k, x_j - x_k \rangle = \|x_j\|^2 + \|x_k\|^2 = 2\pi, \quad j \ne k,$$

and, similarly, $\|y_j - y_k\|^2 = 2\pi$, $j \ne k$, from which we get

$$\|f_j(t) - f_k(t)\| = \begin{cases} \|x_j - x_k\| & \text{if } t \in I_1 \\ \|y_j - y_k\| & \text{if } t \in I_2 \end{cases} = \sqrt{2\pi}, \quad j \ne k.$$

(It already follows that Theorem 4.1 is inapplicable to $\{f_j\}$.)

Given $t \in T$ and $j \in \mathbb{N}$, we have

$$\|f_j(t)\| = \|\mathscr{D}_{x_j,y_j}(t)\| = \begin{cases} \|x_j\| & \text{if } t \in I_1 \\ \|y_j\| & \text{if } t \in I_2 \end{cases} = \sqrt{\pi},$$

and so, conditions (i) in Theorems 4.15 and 4.16 are satisfied.

Let us show that condition (ii) in Theorem 4.15 does not hold. In fact, by (3.29) and (3.30),

$$V_\varepsilon(f_j, T) = \begin{cases} \infty & \text{if } 0 < \varepsilon < \frac{1}{2}\|x_j - y_j\|, \\ 0 & \text{if } \quad \varepsilon \geq \frac{1}{2}\|x_j - y_j\|, \end{cases}$$

where $\|x_j - y_j\|^2 = \langle x_j - y_j, x_j - y_j \rangle = \|x_j\|^2 + \|y_j\|^2 = 2\pi$, i.e., $\|x_j - y_j\| = \sqrt{2\pi}$.

Now, we show that condition (ii) in Theorem 4.16 is satisfied (cf. Example 4.12). By (4.23), for every $x^* \in M^* = M$ and $t \in T$, we have

$$\langle f_j(t), x^* \rangle = \langle \mathscr{D}_{x_j,y_j}(t), x^* \rangle = \langle \mathscr{D}_{1,0}(t)x_j + \mathscr{D}_{0,1}(t)y_j, x^* \rangle$$

$$= \mathscr{D}_{1,0}(t)\langle x_j, x^* \rangle + \mathscr{D}_{0,1}(t)\langle y_j, x^* \rangle = \mathscr{D}_{x'_j, y'_j}(t),$$

where $x'_j = \langle x_j, x^* \rangle$ and $y'_j = \langle y_j, x^* \rangle$. Again, by (3.29) and (3.30),

$$V_\varepsilon(\langle f_j(\cdot), x^* \rangle, T) = V_\varepsilon(\mathscr{D}_{x'_j, y'_j}, T) = \begin{cases} \infty & \text{if } 0 < \varepsilon < \frac{1}{2}|x'_j - y'_j|, \\ 0 & \text{if } \quad \varepsilon \geq \frac{1}{2}|x'_j - y'_j|, \end{cases} \quad (4.24)$$

where $|x'_j - y'_j| = |\langle x_j, x^* \rangle - \langle y_j, x^* \rangle| \to 0$ as $j \to \infty$. Hence, given $\varepsilon > 0$, there is $j_0 = j_0(\varepsilon, x^*) \in \mathbb{N}$ such that $|x'_j - y'_j| \leq 2\varepsilon$ for all $j \geq j_0$, and so, (4.24) implies $V_\varepsilon(\langle f_j(\cdot), x^* \rangle, T) = 0$ for all $j \geq j_0$. Thus,

$$\limsup_{j \to \infty} V_\varepsilon(\langle f_j(\cdot), x^* \rangle, T) \leq \sup_{j \geq j_0} V_\varepsilon(\langle f_j(\cdot), x^* \rangle, T) = 0$$

(i.e., $V_\varepsilon(\langle f_j(\cdot), x^* \rangle, T) \to 0$, $j \to \infty$), which yields condition (ii) in Theorem 4.16.

4.5 Irregular pointwise selection principles

In what follows, we shall be dealing with double sequences $\alpha : \mathbb{N} \times \mathbb{N} \to [0, \infty]$ having the property that $\alpha(j, j) = 0$ for all $j \in \mathbb{N}$ (e.g., (4.25)). The *limit superior* of $\alpha(j, k)$ as $j, k \to \infty$ is defined by

$$\limsup_{j,k \to \infty} \alpha(j, k) = \lim_{n \to \infty} \sup\{\alpha(j, k) : j \geq n \text{ and } k \geq n\}.$$

For a number $\alpha_0 \geq 0$, we say that $\alpha(j, k)$ *converges* to α_0 as $j, k \to \infty$ and write $\lim_{j,k\to\infty} \alpha(j, k) = \alpha_0$ if for every $\eta > 0$ there is $J = J(\eta) \in \mathbb{N}$ such that

$$|\alpha(j, k) - \alpha_0| \leq \eta \quad \text{for all} \quad j \geq J \text{ and } k \geq J \text{ with } j \neq k.$$

The main result of this section is the following *irregular pointwise selection principle* in terms of the approximate variation (see also Example 4.27).

Theorem 4.18 *Suppose $T \subset \mathbb{R}$, $(M, \| \cdot \|)$ is a normed linear space, and $\{f_j\} \subset M^T$ is a pointwise relatively compact sequence of functions such that*

$$\limsup_{j,k\to\infty} V_\varepsilon(f_j - f_k, T) < \infty \quad \text{for all} \quad \varepsilon > 0. \tag{4.25}$$

Then $\{f_j\}$ contains a subsequence which converges pointwise on T.

In order to prove this theorem, we need a lemma.

Lemma 4.19 *Suppose $\varepsilon > 0$, $C > 0$, and a sequence $\{F_j\}_{j=1}^\infty \subset M^T$ of distinct functions are such that*

$$V_\varepsilon(F_j - F_k, T) \leq C \quad \text{for all} \quad j, k \in \mathbb{N}. \tag{4.26}$$

Then, there exists a subsequence $\{F_j^\varepsilon\}_{j=1}^\infty$ of $\{F_j\}_{j=1}^\infty$ and a nondecreasing function $\varphi^\varepsilon : T \to [0, C]$ such that

$$\lim_{j,k\to\infty} V_\varepsilon(F_j^\varepsilon - F_k^\varepsilon, T \cap (-\infty, t]) = \varphi^\varepsilon(t) \quad \text{for all} \quad t \in T. \tag{4.27}$$

Since the proof of Lemma 4.19 is rather lengthy and involves certain ideas from formal logic (Ramsey's Theorem 4.20), for the time being we postpone it until the end of the proof of Theorem 4.18.

Proof of Theorem 4.18 First, we may assume that T is *uncountable*. In fact, if T is (at most) countable, then, by the relative compactness of sets $\{f_j(t) : j \in \mathbb{N}\} \subset M$ for all $t \in T$, we may apply the standard diagonal procedure to extract a subsequence of $\{f_j\}$ which converges pointwise on T. Second, we may assume that all functions in the sequence $\{f_j\}$ are *distinct*. To see this, we argue as follows. If there are only finitely many distinct functions in $\{f_j\}$, then we may choose a constant subsequence of $\{f_j\}$ (which is, clearly, pointwise convergent on T). Otherwise, we may pick a subsequence of $\{f_j\}$ (if necessary) consisting of distinct functions.

Given $\varepsilon > 0$, we set (cf. (4.25))

$$C(\varepsilon) = 1 + \limsup_{j,k\to\infty} V_\varepsilon(f_j - f_k, T) < \infty.$$

So, there is $j_0(\varepsilon) \in \mathbb{N}$ such that

$$V_\varepsilon(f_j - f_k, T) \le C(\varepsilon) \quad \text{for all} \quad j \ge j_0(\varepsilon) \text{ and } k \ge j_0(\varepsilon). \tag{4.28}$$

Let $\{\varepsilon_n\}_{n=1}^\infty \subset (0, \infty)$ be a decreasing sequence such that $\varepsilon_n \to 0$ as $n \to \infty$.
    We divide the rest of the proof into two main steps for clarity.

Step 1 There is a subsequence of $\{f_j\}$, again denoted by $\{f_j\}$, and for each $n \in \mathbb{N}$ there is a nondecreasing function $\varphi_n : T \to [0, C(\varepsilon_n)]$ such that

$$\lim_{j,k \to \infty} V_{\varepsilon_n}(f_j - f_k, T \cap (-\infty, t]) = \varphi_n(t) \quad \text{for all} \quad t \in T. \tag{4.29}$$

In order to prove (4.29), we apply Lemma 4.19, induction, and the diagonal procedure. Setting $\varepsilon = \varepsilon_1$, $C = C(\varepsilon_1)$, and $F_j = f_{J_0(j)}$ with $J_0(j) = j_0(\varepsilon_1) + j - 1$, $j \in \mathbb{N}$, we find that condition (4.28) implies (4.26), and so, by Lemma 4.19, there are a subsequence $\{J_1(j)\}_{j=1}^\infty$ of $\{J_0(j)\}_{j=1}^\infty = \{j\}_{j=j_0(\varepsilon_1)}^\infty$ and a nondecreasing function $\varphi_1 = \varphi^{\varepsilon_1} : T \to [0, C(\varepsilon_1)]$ such that

$$\lim_{j,k \to \infty} V_{\varepsilon_1}(f_{J_1(j)} - f_{J_1(k)}, T \cap (-\infty, t]) = \varphi_1(t) \quad \text{for all} \quad t \in T.$$

Let $j_1 \in \mathbb{N}$ be the least number such that $J_1(j_1) \ge j_0(\varepsilon_2)$. Inductively, suppose $n \in \mathbb{N}$, $n \ge 2$, and a subsequence $\{J_{n-1}(j)\}_{j=1}^\infty$ of $\{j\}_{j=j_0(\varepsilon_1)}^\infty$ and a number $j_{n-1} \in \mathbb{N}$ with $J_{n-1}(j_{n-1}) \ge j_0(\varepsilon_n)$ are already chosen. To apply Lemma 4.19 once again, we set $\varepsilon = \varepsilon_n$, $C = C(\varepsilon_n)$, and $F_j = f_{J(j)}$ with $J(j) = J_{n-1}(j_{n-1} + j - 1)$, $j \in \mathbb{N}$. Since, for every $j \in \mathbb{N}$, we have $J(j) \ge J_{n-1}(j_{n-1}) \ge j_0(\varepsilon_n)$, we get, by (4.28),

$$V_{\varepsilon_n}(F_j - F_k, T) \le C(\varepsilon_n) \quad \text{for all} \quad j, k \in \mathbb{N}.$$

By Lemma 4.19, there are a subsequence $\{J_n(j)\}_{j=1}^\infty$ of the sequence $\{J(j)\}_{j=1}^\infty$, (more explicitly) the latter being equal to $\{J_{n-1}(j)\}_{j=j_{n-1}}^\infty$, and a nondecreasing function $\varphi_n = \varphi^{\varepsilon_n} : T \to [0, C(\varepsilon_n)]$ such that

$$\lim_{j,k \to \infty} V_{\varepsilon_n}(f_{J_n(j)} - f_{J_n(k)}, T \cap (-\infty, t]) = \varphi_n(t) \quad \text{for all} \quad t \in T. \tag{4.30}$$

We assert that the diagonal subsequence $\{f_{J_j(j)}\}_{j=1}^\infty$ of $\{f_j\}$, again denoted by $\{f_j\}$, satisfies (4.29) for all $n \in \mathbb{N}$. In order to see this, let us fix $n \in \mathbb{N}$ and $t \in T$. By (4.30), given $\eta > 0$, there is a number $J^0 = J^0(\eta, n, t) \in \mathbb{N}$ such that if $j', k' \ge J^0$, $j' \ne k'$, we have

$$\left| V_{\varepsilon_n}(f_{J_n(j')} - f_{J_n(k')}, T \cap (-\infty, t]) - \varphi_n(t) \right| \le \eta. \tag{4.31}$$

Since $\{J_j(j)\}_{j=n}^\infty$ is a subsequence of $\{J_n(j)\}_{j=1}^\infty$, there is a strictly increasing natural sequence $q : \mathbb{N} \to \mathbb{N}$ such that $J_j(j) = J_n(q(j))$ for all $j \ge n$. Define $J^* = \max\{n, J^0\}$. Now, for arbitrary $j, k \ge J^*$, $j \ne k$, we set $j' = q(j)$ and

$k' = q(k)$. Since $j, k \geq J^* \geq n$, we find $J_j(j) = J_n(j')$ and $J_k(k) = J_n(k')$, where $j' \neq k'$, $j' = q(j) \geq j \geq J^* \geq J^0$ and, similarly, $k' \geq J^0$. It follows from (4.31) that

$$\left| V_{\varepsilon_n}(f_{J_j(j)} - f_{J_k(k)}, T \cap (-\infty, t]) - \varphi_n(t) \right| \leq \eta.$$

which proves our assertion.

Step 2 Let Q denote an at most countable dense subset of T. Clearly, Q contains every point of T which is not a limit point for T. Since, for any $n \in \mathbb{N}$, the function φ_n from (4.29) is nondecreasing on T, the set $Q_n \subset T$ of its points of discontinuity is at most countable. We set $S = Q \cup \bigcup_{n=1}^{\infty} Q_n$. The set S is an at most countable dense subset of T and has the property:

$$\text{for each } n \in \mathbb{N}, \text{ the function } \varphi_n \text{ is continuous on } T \setminus S. \tag{4.32}$$

By the relative compactness of the set $\{f_j(t) : j \in \mathbb{N}\}$ for all $t \in T$ and at most countability of $S \subset T$, we may assume (applying the diagonal procedure and passing to a subsequence of $\{f_j\}$ if necessary) that, for every $s \in S$, $f_j(s)$ converges in M as $j \to \infty$ to a point of M denoted by $f(s)$ (hence $f : S \to M$).

It remains to show that the sequence $\{f_j(t)\}_{j=1}^{\infty}$ is Cauchy in M for every point $t \in T \setminus S$. In fact, this and the relative compactness of $\{f_j(t) : j \in \mathbb{N}\}$ imply the convergence of $f_j(t)$ as $j \to \infty$ to a point of M denoted by $f(t)$. In other words, f_j converges pointwise on T to the function $f : T = S \cup (T \setminus S) \to M$.

Let $t \in T \setminus S$ and $\eta > 0$ be arbitrary. Since $\varepsilon_n \to 0$ as $n \to \infty$, choose and fix $n = n(\eta) \in \mathbb{N}$ such that $\varepsilon_n \leq \eta$. The definition of S implies that t is a limit point for T and a point of continuity of φ_n, and so, by the density of S in T, there is $s = s(n, t) \in S$ such that $|\varphi_n(t) - \varphi_n(s)| \leq \eta$. Property (4.29) yields the existence of $j^1 = j^1(\eta, n, t, s) \in \mathbb{N}$ such that if $j, k \geq j^1$, $j \neq k$,

$$\left| V_{\varepsilon_n}(f_j - f_k, T \cap (-\infty, \tau]) - \varphi_n(\tau) \right| \leq \eta \quad \text{for} \quad \tau = t \text{ and } \tau = s.$$

Suppose $s < t$ (the case when $s > t$ is treated similarly). Applying Lemma 2.8 (with T replaced by $T \cap (-\infty, t]$, T_1—by $T \cap (-\infty, s]$, and T_2—by $T \cap [s, t]$), we get

$$V_{\varepsilon_n}(f_j - f_k, T \cap [s, t]) \leq V_{\varepsilon_n}(f_j - f_k, T \cap (-\infty, t]) - V_{\varepsilon_n}(f_j - f_k, T \cap (-\infty, s])$$

$$\leq |V_{\varepsilon_n}(f_j - f_k, T \cap (-\infty, t]) - \varphi_n(t)| + |\varphi_n(t) - \varphi_n(s)|$$

$$+ |\varphi_n(s) - V_{\varepsilon_n}(f_j - f_k, T \cap (-\infty, s])|$$

$$\leq \eta + \eta + \eta = 3\eta \quad \text{for all} \quad j, k \geq j^1 \text{ with } j \neq k.$$

Now, given $j, k \geq j^1$, $j \neq k$, by the definition of $V_{\varepsilon_n}(f_j - f_k, T \cap [s, t])$, there is a function $g_{j,k} \in BV(T \cap [s, t]; M)$, also depending on η, n, t, and s, such that

$$\|(f_j - f_k) - g_{j,k}\|_{\infty, T \cap [s,t]} \leq \varepsilon_n$$

and

$$V(g_{j,k}, T \cap [s, t]) \leq V_{\varepsilon_n}(f_j - f_k, T \cap [s, t]) + \eta.$$

These inequalities and (2.2) imply, for all $j, k \geq j^1$ with $j \neq k$,

$$\|(f_j - f_k)(s) - (f_j - f_k)(t)\| \leq \|g_{j,k}(s) - g_{j,k}(t)\| + 2\|(f_j - f_k) - g_{j,k}\|_{\infty, T \cap [s,t]}$$

$$\leq V(g_{j,k}, T \cap [s, t]) + 2\varepsilon_n \leq (3\eta + \eta) + 2\eta = 6\eta.$$

Since the sequence $\{f_j(s)\}_{j=1}^{\infty}$ is convergent in M, it is a Cauchy sequence, and so, there is $j^2 = j^2(\eta, s) \in \mathbb{N}$ such that $\|f_j(s) - f_k(s)\| \leq \eta$ for all $j, k \geq j^2$. It follows that $j^3 = \max\{j^1, j^2\}$ depends only on η (and t), and we have

$$\|f_j(t) - f_k(t)\| \leq \|(f_j - f_k)(t) - (f_j - f_k)(s)\| + \|(f_j - f_k)(s)\|$$

$$\leq 6\eta + \eta = 7\eta \quad \text{for all} \quad j, k \geq j^3.$$

Thus, $\{f_j(t)\}_{j=1}^{\infty}$ is a Cauchy sequence in M, which completes the proof.   □

Various remarks and examples concerning Theorem 4.18 follow after the proof of Lemma 4.19.

Now we turn to the proof of Lemma 4.19. We need Ramsey's Theorem from formal logic [63, Theorem A], which we are going to recall now.

Let Γ be a set, $n \in \mathbb{N}$, and $\gamma_1, \gamma_2, \ldots, \gamma_n$ be (pairwise) distinct elements of Γ. The (non-ordered) collection $\{\gamma_1, \gamma_2, \ldots, \gamma_n\}$ is said to be an *n-combination* of elements of Γ (note that an n-combination may be generated by $n!$ different injective functions $\gamma : \{1, 2, \ldots, n\} \to \Gamma$ with $\gamma_i = \gamma(i)$ for all $i = 1, 2, \ldots, n$). We denote by $\Gamma[n]$ the family of all n-combinations of elements of Γ.

Theorem 4.20 (Ramsey [63]) *Suppose Γ is an infinite set, $n, m \in \mathbb{N}$, and $\Gamma[n] = \bigcup_{i=1}^{m} G_i$ is a disjoint union of m nonempty sets $G_i \subset \Gamma[n]$. Then, under the Axiom of Choice, there are an infinite set $\Delta \subset \Gamma$ and $i_0 \in \{1, 2, \ldots, m\}$ such that $\Delta[n] \subset G_{i_0}$.*

This theorem will be applied several times in the proof of Lemma 4.19 with Γ a subset of $\{F_j : j \in \mathbb{N}\}$ and $n = m = 2$.

The application of Ramsey's Theorem in the context of pointwise selection principles was initiated by Schrader [68] and later on was extended by several authors (Di Piazza and Maniscalco [41], Maniscalco [54], Chistyakov and Maniscalco [33],

Chistyakov, Maniscalco and Tretyachenko [34], Chistyakov and Tretyachenko [38]) for real- and metric space-valued functions of one and several real variables.

Proof of Lemma 4.19 We divide the proof into three steps.

Step 1 Let us show that for every $t \in T$ there is a subsequence $\{F_j^{(t)}\}_{j=1}^{\infty}$ of $\{F_j\}_{j=1}^{\infty}$, depending on t and ε, such that the double limit

$$\lim_{j,k \to \infty} V_\varepsilon(F_j^{(t)} - F_k^{(t)}, T \cap (-\infty, t]) \quad \text{exists in} \quad [0, C] \tag{4.33}$$

(clearly, the sequence $\{F_j^{(t)}\}_{j=1}^{\infty}$ satisfies the uniform estimate (4.26)).

Given $t \in T$, for the sake brevity, we set $T_t^- = T \cap (-\infty, t]$. By Lemma 2.2(b) and (4.26), we have

$$0 \le V_\varepsilon(F_j - F_k, T_t^-) \le V_\varepsilon(F_j - F_k, T) \le C \quad \text{for all} \quad j, k \in \mathbb{N}.$$

In order to apply Theorem 4.20, we set $\Gamma = \{F_j : j \in \mathbb{N}\}$, $c_0 = C/2$, and denote by G_1 the set of those pairs $\{F_j, F_k\}$ with $j, k \in \mathbb{N}$, $j \ne k$, for which $V_\varepsilon(F_j - F_k, T_t^-) \in [0, c_0)$, and by G_2—the set of all pairs $\{F_j, F_k\}$ with $j, k \in \mathbb{N}$, $j \ne k$, having the property that $V_\varepsilon(F_j - F_k, T_t^-) \in [c_0, C]$. Clearly, $\Gamma[2] = G_1 \cup G_2$ and $G_1 \cap G_2 = \varnothing$. If G_1 and G_2 are both nonempty, then, by Theorem 4.20, there is a subsequence $\{F_j^1\}_{j=1}^{\infty}$ of $\{F_j\}_{j=1}^{\infty}$ (cf. Remark 4.21) such that either

(i$_1$) $\{F_j^1, F_k^1\} \in G_1$ for all $j, k \in \mathbb{N}$, $j \ne k$, or
(ii$_1$) $\{F_j^1, F_k^1\} \in G_2$ for all $j, k \in \mathbb{N}$, $j \ne k$.

In the case when $G_1 \ne \varnothing$ and (i$_1$) holds, or $G_2 = \varnothing$, we set $[a_1, b_1] = [0, c_0]$, while if $G_2 \ne \varnothing$ and (ii$_1$) holds, or $G_1 = \varnothing$, we set $[a_1, b_1] = [c_0, C]$.

Inductively, assume that $p \in \mathbb{N}$, $p \ge 2$, and a subsequence $\{F_j^{p-1}\}_{j=1}^{\infty}$ of $\{F_j\}_{j=1}^{\infty}$ and an interval $[a_{p-1}, b_{p-1}] \subset [0, C]$ such that

$$V_\varepsilon(F_j^{p-1} - F_k^{p-1}, T_t^-) \in [a_{p-1}, b_{p-1}] \quad \text{for all} \quad j, k \in \mathbb{N}, \; j \ne k,$$

are already chosen. To apply Theorem 4.20, we set $\Gamma = \{F_j^{p-1} : j \in \mathbb{N}\}$, define $c_{p-1} = \frac{1}{2}(a_{p-1} + b_{p-1})$, and denote by G_1 the set of all pairs $\{F_j^{p-1}, F_k^{p-1}\}$ with $j, k \in \mathbb{N}$, $j \ne k$, such that $V_\varepsilon(F_j^{p-1} - F_k^{p-1}, T_t^-) \in [a_{p-1}, c_{p-1})$, and by G_2—the set of all pairs $\{F_j^{p-1}, F_k^{p-1}\}$ with $j, k \in \mathbb{N}$, $j \ne k$, for which $V_\varepsilon(F_j^{p-1} - F_k^{p-1}, T_t^-) \in [c_{p-1}, b_{p-1}]$. We have the union $\Gamma[2] = G_1 \cup G_2$ of disjoint sets. If G_1 and G_2 are both nonempty, then, by Ramsey's Theorem, there is a subsequence $\{F_j^p\}_{j=1}^{\infty}$ of $\{F_j^{p-1}\}_{j=1}^{\infty}$ such that either

(i$_p$) $\{F_j^p, F_k^p\} \in G_1$ for all $j, k \in \mathbb{N}$, $j \ne k$, or
(ii$_p$) $\{F_j^p, F_k^p\} \in G_2$ for all $j, k \in \mathbb{N}$, $j \ne k$.

If $G_1 \neq \varnothing$ and (i_p) holds, or $G_2 = \varnothing$, we set $[a_p, b_p] = [a_{p-1}, c_{p-1}]$, while if $G_2 \neq \varnothing$ and (ii_p) holds, or $G_1 = \varnothing$, we set $[a_p, b_p] = [c_{p-1}, b_{p-1}]$.

In this way for each $p \in \mathbb{N}$ we have nested intervals $[a_p, b_p] \subset [a_{p-1}, b_{p-1}]$ in $[a_0, b_0] = [0, C]$ with $b_p - a_p = C/2^p$ and a subsequence $\{F_j^p\}_{j=1}^\infty$ of $\{F_j^{p-1}\}_{j=1}^\infty$ (where $F_j^0 = F_j$, $j \in \mathbb{N}$) such that

$$V_\varepsilon(F_j^p - F_k^p, T_t^-) \in [a_p, b_p] \quad \text{for all} \quad j, k \in \mathbb{N}, \ j \neq k.$$

Let $\ell \in [0, C]$ be the common limit of a_p and b_p as $p \to \infty$ (note that ℓ depends on t and ε). Denoting the diagonal sequence $\{F_j^j\}_{j=1}^\infty$ by $\{F_j^{(t)}\}_{j=1}^\infty$ we infer that the limit in (4.33) is equal to ℓ. In fact, given $\eta > 0$, there is $p(\eta) \in \mathbb{N}$ such that $a_{p(\eta)}, b_{p(\eta)} \in [\ell - \eta, \ell + \eta]$ and, since $\{F_j^{(t)}\}_{j=p(\eta)}^\infty$ is a subsequence of $\{F_j^{p(\eta)}\}_{j=1}^\infty$, we find, for all $j, k \geq p(\eta)$ with $j \neq k$, that

$$V_\varepsilon(F_j^{(t)} - F_k^{(t)}, T_t^-) \in [a_{p(\eta)}, b_{p(\eta)}] \subset [\ell - \eta, \ell + \eta].$$

Step 2 Given a set $A \subset \mathbb{R}$, we denote by $\overline{A}$ its closure in $\mathbb{R}$.

Let Q be an at most countable dense subset of T (hence $Q \subset T \subset \overline{Q}$). The set

$$T_L = \{t \in T : T \cap (t - \delta, t) = \varnothing \text{ for some } \delta > 0\}$$

of points from T, which are isolated from the left for T, is at most countable, and the same is true for the set

$$T_R = \{t \in T : T \cap (t, t + \delta) = \varnothing \text{ for some } \delta > 0\}$$

of points from T isolated from the right for T. Clearly, $T_L \cap T_R \subset Q$, and the set $Z = Q \cup T_L \cup T_R$ is an at most countable dense subset of T.

We assert that there are a subsequence $\{F_j^*\}_{j=1}^\infty$ of $\{F_j\}_{j=1}^\infty$ and a nondecreasing function $\varphi : Z \to [0, C]$ (both depending on ε) such that

$$\lim_{j,k\to\infty} V_\varepsilon(F_j^* - F_k^*, T \cap (-\infty, s]) = \varphi(s) \quad \text{for all} \quad s \in Z. \tag{4.34}$$

With no loss of generality, we may assume that $Z = \{s_p\}_{p=1}^\infty$. By Step 1, there are a subsequence $\{F_j^{(s_1)}\}_{j=1}^\infty$ of $\{F_j\}_{j=1}^\infty$, denoted by $\{F_j^{(1)}\}_{j=1}^\infty$, and a number from $[0, C]$, denoted by $\varphi(s_1)$, such that

$$\lim_{j,k\to\infty} V_\varepsilon(F_j^{(1)} - F_k^{(1)}, T \cap (-\infty, s_1]) = \varphi(s_1).$$

Inductively, if $p \in \mathbb{N}$, $p \geq 2$, and a subsequence $\{F_j^{(p-1)}\}_{j=1}^\infty$ of $\{F_j\}_{j=1}^\infty$ is already chosen, we apply Step 1 once again to pick a subsequence $\{F_j^{(p)}\}_{j=1}^\infty$ of $\{F_j^{(p-1)}\}_{j=1}^\infty$ and a number $\varphi(s_p) \in [0, C]$ such that

$$\lim_{j,k\to\infty} V_\varepsilon(F_j^{(p)} - F_k^{(p)}, T \cap (-\infty, s_p]) = \varphi(s_p).$$

Denoting by $\{F_j^*\}_{j=1}^\infty$ the diagonal subsequence $\{F_j^{(j)}\}_{j=1}^\infty$ of $\{F_j\}_{j=1}^\infty$, we establish (4.34). It remains to note that, by Lemma 2.2(b), the function $\varphi : Z \to [0, C]$, defined by the left-hand side of (4.34), is nondecreasing on Z.

Step 3 In this step, we finish the proof of (4.27). Applying Saks' idea [67, Chapter 7, Section 4, Lemma (4.1)], we extend the function φ, defined by (4.34), from the set Z to the whole $\mathbb{R}$ as follows: given $t \in \mathbb{R}$,

$$\widetilde{\varphi}(t) = \sup\{\varphi(s) : s \in Z \cap (-\infty, t]\} \quad \text{if} \quad Z \cap (-\infty, t] \neq \varnothing$$

and

$$\widetilde{\varphi}(t) = \inf\{\varphi(s) : s \in Z\} \quad \text{otherwise.}$$

Clearly, $\widetilde{\varphi} : \mathbb{R} \to [0, \infty)$ is nondecreasing and $\widetilde{\varphi}(\mathbb{R}) \subset \overline{\varphi(Z)} \subset [0, C]$. Therefore, the set $D \subset \mathbb{R}$ of points of discontinuity of $\widetilde{\varphi}$ is at most countable.

Let us show that if $\{F_j^*\}_{j=1}^\infty$ is the sequence from (4.34), then

$$\lim_{j,k\to\infty} V_\varepsilon(F_j^* - F_k^*, T \cap (-\infty, t]) = \widetilde{\varphi}(t) \quad \text{for all} \quad t \in T \setminus D. \tag{4.35}$$

By virtue of (4.34), we may assume that $t \in T \setminus (D \cup Z)$. Let $\eta > 0$ be fixed. Since t is a point of continuity of $\widetilde{\varphi}$, there is $\delta = \delta(\eta) > 0$ such that

$$\widetilde{\varphi}(s) \in [\widetilde{\varphi}(t) - \eta, \widetilde{\varphi}(t) + \eta] \text{ for all } s \in \mathbb{R} \text{ such that } |s - t| \leq \delta. \tag{4.36}$$

Since $T \subset \overline{Z}$ and $t \notin T_L$, we find $\overline{Z} \cap (t - \delta, t) \supset T \cap (t - \delta, t) \neq \varnothing$, and so, there is $s' \in Z$ with $t - \delta < s' < t$. By (4.34), there is $j^1 = j^1(\eta) \in \mathbb{N}$ such that, for all $j, k \geq j^1$, $j \neq k$,

$$V_\varepsilon(F_j^* - F_k^*, T \cap (-\infty, s']) \in [\varphi(s') - \eta, \varphi(s') + \eta]. \tag{4.37}$$

Similarly, $t \notin T_R$ implies the existence of $s'' \in Z$ with $t < s'' < t + \delta$, and so, by (4.34), for some $j^2 = j^2(\eta) \in \mathbb{N}$, we have, for all $j, k \geq j^2$, $j \neq k$,

$$V_\varepsilon(F_j^* - F_k^*, T \cap (-\infty, s'']) \in [\varphi(s'') - \eta, \varphi(s'') + \eta]. \tag{4.38}$$

Since $s' < t < s''$, $T \cap (-\infty, s'] \subset T \cap (-\infty, t] \subset T \cap (-\infty, s'']$, and so, by Lemma 2.2(b), we get, for all $j, k \in \mathbb{N}$,

$$V_\varepsilon(F_j^* - F_k^*, T \cap (-\infty, s']) \leq V_\varepsilon(F_j^* - F_k^*, T \cap (-\infty, t])$$
$$\leq V_\varepsilon(F_j^* - F_k^*, T \cap (-\infty, s'']).$$

Setting $j^3 = \max\{j^1, j^2\}$ and noting that $\widetilde{\varphi}(s') = \varphi(s')$ and $\widetilde{\varphi}(s'') = \varphi(s'')$, we find, from (4.37), (4.38), and (4.36), that

$$V_\varepsilon(F_j^* - F_k^*, T \cap (-\infty, t])$$

$$\in \left[V_\varepsilon(F_j^* - F_k^*, T \cap (-\infty, s']), \; V_\varepsilon(F_j^* - F_k^*, T \cap (-\infty, s'']) \right]$$

$$\subset [\varphi(s') - \eta, \varphi(s'') + \eta] = [\widetilde{\varphi}(s') - \eta, \widetilde{\varphi}(s'') + \eta]$$

$$\subset [\widetilde{\varphi}(t) - 2\eta, \widetilde{\varphi}(t) + 2\eta] \quad \text{for all} \quad j, k \geq j^3, \; j \neq k,$$

which proves (4.35).

Finally, we note that $T = (T \setminus D) \cup (T \cap D)$, where $T \cap D$ is at most countable. Furthermore, being a subsequence of the original sequence $\{F_j\}_{j=1}^\infty$, the sequence $\{F_j^*\}_{j=1}^\infty$ from (4.34) and (4.35) satisfies the uniform estimate (4.26). So, arguing as in Step 2 with Z replaced by $T \cap D$, we obtain a subsequence of $\{F_j^*\}_{j=1}^\infty$, denoted by $\{F_j^\varepsilon\}_{j=1}^\infty$, and a nondecreasing function $\psi : T \cap D \to [0, C]$ such that

$$\lim_{j,k\to\infty} V_\varepsilon(F_j^\varepsilon - F_k^\varepsilon, T \cap (-\infty, t]) = \psi(t) \quad \text{for all} \quad t \in T \cap D. \tag{4.39}$$

We define the desired function $\varphi^\varepsilon : T \to [0, C]$ by $\varphi^\varepsilon(t) = \widetilde{\varphi}(t)$ if $t \in T \setminus D$ and $\varphi^\varepsilon(t) = \psi(t)$ if $t \in T \cap D$. Now, it follows from (4.35) and (4.39) that equality (4.27) holds, where, in view of Lemma 2.2(b), the function φ^ε is nondecreasing on T.

This completes the proof of Lemma 4.19. $\square$

Remark 4.21 Here we present more details on the existence of the subsequence $\{F_j^1\}_{j=1}^\infty$ of $\{F_j\}_{j=1}^\infty$ after the first application of Ramsey's Theorem (cf. p. 73). By Theorem 4.20, there is an infinite set $\Delta \subset \Gamma = \{F_j : j \in \mathbb{N}\}$ such that either $\Delta[2] \subset G_1$ or $\Delta[2] \subset G_2$. We infer that

$$\Delta = \{F_{q(n)} : n \in \mathbb{N}\} \text{ for some strictly increasing sequence } q : \mathbb{N} \to \mathbb{N}, \tag{4.40}$$

and, setting $F_j^1 = F_{q(j)}$ for $j \in \mathbb{N}$, we have $\Delta[2] = \{\{F_j^1, F_k^1\} : j, k \in \mathbb{N}, \; j \neq k\}$.

Since the set $\mathbb{N}$ of natural numbers is well-ordered (i.e., every nonempty subset of $\mathbb{N}$ has the minimal element), the sequence $q : \mathbb{N} \to \mathbb{N}$ can be defined as follows: $q(1) = \min\{j \in \mathbb{N} : F_j \in \Delta\}$, and, inductively, if $n \in \mathbb{N}$, $n \geq 2$, and natural numbers $q(1) < q(2) < \ldots < q(n-1)$ are already defined, we set

$$q(n) = \min\{j \in \mathbb{N} \setminus \{q(1), q(2), \ldots, q(n-1)\} : F_j \in \Delta\}. \tag{4.41}$$

The sequence q is strictly increasing: if $n \in \mathbb{N}$ and $j \in \mathbb{N} \setminus \{q(1), q(2), \ldots, q(n)\}$ is such that $F_j \in \Delta$, then $j \neq q(n)$, and since $j \in \mathbb{N} \setminus \{q(1), q(2), \ldots, q(n-1)\}$, we have, by (4.41), $j \geq q(n)$, i.e., $j > q(n)$; by the arbitrariness of j as above and (4.41) (for $n + 1$ in place of n), we get $q(n + 1) > q(n)$. Clearly, $q(n) \geq n$.

Let us verify the equality in (4.40). The inclusion ($\supset$) is clear from (4.41). To see that inclusion ($\subset$) holds, let $F \in \Delta$, so that $\Delta \subset \Gamma$ implies $F = F_{j_0}$ for some $j_0 \in \mathbb{N}$. We have $q(j_0) \geq j_0$, and since $F_{j_0} \in \Delta$, $j_0 \geq q(1)$. Hence $q(1) \leq j_0 \leq q(j_0)$. We claim that there is $1 \leq n_0 \leq j_0$ such that $q(n_0) = j_0$ (this implies $F = F_{j_0} = F_{q(n_0)} \in \{F_{q(n)} : n \in \mathbb{N}\}$ and establishes ($\subset$)). By contradiction, if $q(n) \neq j_0$ for all $n = 1, 2, \ldots, j_0$, then j_0 belongs to the set $\{j \in \mathbb{N} \setminus \{q(1), q(2), \ldots, q(j_0)\} : F_j \in \Delta\}$, and so, by (4.41), $q(j_0 + 1) \leq j_0$, which contradicts $q(j_0 + 1) > q(j_0) \geq j_0$.

Remark 4.22 If $(M, \|\cdot\|)$ is a *finite-dimensional* normed linear space, the condition of relative compactness of sets $\{f_j(t) : j \in \mathbb{N}\}$ at all points $t \in T$ in Theorem 4.18 can be lightened to the condition $\sup_{j \in \mathbb{N}} \|f_j(t_0)\| \equiv C_0 < \infty$ for some $t_0 \in T$. In fact, by Lemma 2.5(b) and (4.28) with fixed $\varepsilon_0 > 0$ and $j_0 \equiv j_0(\varepsilon_0)$, we get

$$|(f_j - f_{j_0})(T)| \leq V_{\varepsilon_0}(f_j - f_{j_0}, T) + 2\varepsilon_0 \leq C(\varepsilon_0) + 2\varepsilon_0 \quad \text{for all} \quad j \geq j_0.$$

Hence, given $t \in T$, we find

$$\|f_j(t)\| \leq \|(f_j - f_{j_0})(t) - (f_j - f_{j_0})(t_0)\| + \|f_{j_0}(t)\| + \|f_j(t_0)\| + \|f_{j_0}(t_0)\|$$

$$\leq (C(\varepsilon_0) + 2\varepsilon_0) + \|f_{j_0}(t)\| + 2C_0 \quad \text{for all} \quad j \geq j_0,$$

and so, the set $\{f_j(t) : j \in \mathbb{N}\}$ is relatively compact in M.

Remark 4.23 Under the assumptions on T and M from Theorem 4.18, if a sequence $\{f_j\} \subset M^T$ converges *uniformly* on T to a function $f \in M^T$, then

$$\lim_{j,k \to \infty} V_\varepsilon(f_j - f_k, T) = 0 \quad \text{for all} \quad \varepsilon > 0, \tag{4.42}$$

i.e., condition (4.25) is *necessary*. In fact, given $\varepsilon > 0$, there is $j_0 = j_0(\varepsilon) \in \mathbb{N}$ such that $\|f_j - f_k\|_{\infty,T} \leq \varepsilon$ for all $j, k \geq j_0(\varepsilon)$. Since the zero function 0 on T is constant, we get $V_\varepsilon(f_j - f_k, T) \leq V(0, T) = 0$ for all $j, k \geq j_0(\varepsilon)$.

Remark 4.24 In Example 4.26, we show that condition (4.25) is *not necessary* for the pointwise convergence of $\{f_j\}$ to f. However, it is "almost necessary" in the following sense (cf. Remark 4.6(b)). Let $T \subset \mathbb{R}$ be a measurable set with *finite* Lebesgue measure $\mathscr{L}(T)$ and $\{f_j\} \subset M^T$ be a sequence of measurable functions which converges pointwise or almost everywhere on T to a function $f \in M^T$. Egorov's Theorem implies that for every $\eta > 0$ there is a measurable set $T_\eta \subset T$ such that $\mathscr{L}(T \setminus T_\eta) \leq \eta$ and $f_j \rightrightarrows f$ on T_η. By (4.42), we get

$$\lim_{j,k \to \infty} V_\varepsilon(f_j - f_k, T_\eta) = 0 \quad \text{for all} \quad \varepsilon > 0.$$

Applying Theorem 4.18 and the diagonal procedure, we get the following.

Theorem 4.25 *Under the assumptions of Theorem 4.18, if a sequence of functions* $\{f_j\} \subset M^T$ *is such that, for all* $\varepsilon > 0$,

$$\limsup_{j,k\to\infty} V_\varepsilon(f_j - f_k, T \setminus E) < \infty \text{ for an at most countable } E \subset T$$

or

$$\limsup_{j,k\to\infty} V_\varepsilon(f_j - f_k, T \cap [a, b]) < \infty \text{ for all } a, b \in T, \ a \leq b,$$

then $\{f_j\}$ *contains a subsequence which converges pointwise on* T.

Example 4.26 Condition (4.25) is *not necessary* for the pointwise convergence even if all functions in the sequence $\{f_j\}$ are regulated. To see this, let $\{f_j\} \subset M^T$ be the sequence from Example 3.9, where $T = I = [0, 1]$ and $(M, \|\cdot\|)$ is a normed linear space. First, note that

$$\limsup_{j,k\to\infty} V_\varepsilon(f_j - f_k, T) \geq \limsup_{j\to\infty} V_\varepsilon(f_j - f_{j+1}, T) \quad \text{for all} \quad \varepsilon > 0. \qquad (4.43)$$

Let us fix $j \in \mathbb{N}$ and set $t_k = k/(j+1)!, k = 0, 1, \ldots, (j+1)!$, so that $f_{j+1}(t_k) = x$ for all such k. We have $f_j(t_k) = x$ if and only if $j!t_k$ is an integer, i.e., $k = n(j+1)$ with $n = 0, 1, \ldots, j!$. It follows that $(f_j - f_{j+1})(t) = y - x$ if $t = t_k$ for those $k \in \{0, 1, \ldots, (j+1)!\}$, for which $k \neq n(j+1)$ for all $n \in \{0, 1, \ldots, j!\}$ (and, in particular, for $k = 1, 2, \ldots, j$); in the remaining cases of $t \in T$ we have $(f_j - f_{j+1})(t) = 0$. If $s_k = \frac{1}{2}(t_{k-1} + t_k) = (k - \frac{1}{2})/(j+1)!, k = 1, 2, \ldots, (j+1)!$, we get a partition of the interval $T = [0, 1]$ of the form

$$0 = t_0 < s_1 < t_1 < s_2 < t_2 < \cdots < s_{(j+1)!} < t_{(j+1)!} = 1,$$

and $f_j(s_k) = f_{j+1}(s_k) = y$ for all $k = 1, 2, \ldots, j$. Now, let $0 < \varepsilon < \frac{1}{2}\|x - y\|$, and a function $g \in M^T$ be arbitrary such that $\|(f_j - f_{j+1}) - g\|_{\infty,T} \leq \varepsilon$. By (2.2), we find

$$V(g, T) \geq \sum_{k=1}^{(j+1)!} \|g(t_k) - g(s_k)\| \geq \sum_{k=1}^{j} (\|(f_j - f_{j+1})(t_k) - (f_j - f_{j+1})(s_k)\| - 2\varepsilon)$$

$$= (\|y - x\| - 2\varepsilon)j,$$

and so, by (2.6), $V_\varepsilon(f_j - f_{j+1}, T) \geq (\|y - x\| - 2\varepsilon)j$. Hence, (4.43) implies

$$\limsup_{j,k\to\infty} V_\varepsilon(f_j - f_k, T) = \infty \quad \text{for all} \quad 0 < \varepsilon < \frac{1}{2}\|x - y\|.$$

Example 4.27 Under the assumptions of Theorem 4.18, we cannot infer that the limit function f of an extracted subsequence of $\{f_j\}$ is a *regulated* function (this is the reason to term this theorem an *irregular* selection principle).

Let $T = [a, b]$ be a closed interval, $(M, \|\cdot\|)$ be a normed linear space, $x, y \in M$, $x \neq y$, and $\alpha_j = 1 + (1/j)$, $j \in \mathbb{N}$ (cf. Example 3.7). The sequence of Dirichlet's functions $f_j = \alpha_j \mathscr{D}_{x,y} = \mathscr{D}_{\alpha_j x, \alpha_j y}$, $j \in \mathbb{N}$, converges *uniformly* on T to the Dirichlet function $f = \mathscr{D}_{x,y}$, which is nonregulated. By virtue of (4.42), Theorem 4.18 can be applied to the sequence $\{f_j\}$. On the other hand, Example 3.7 shows that $\{f_j\}$ does not satisfy condition (4.2), and so, Theorem 4.1 is inapplicable.

Sometimes it is more appropriate to apply Theorem 4.18 in the form of Theorem 4.25. Let $\{\beta_j\}_{j=1}^{\infty} \subset \mathbb{R}$ be a bounded sequence (not necessarily convergent). Formally, Theorem 4.18 cannot be applied to the sequence $f_j = \beta_j \mathscr{D}_{x,y}$, $j \in \mathbb{N}$, on $T = [a, b]$ (e.g., with $\beta_j = (-1)^j$ or $\beta_j = (-1)^j + (1/j)$). However, for every $j \in \mathbb{N}$ the restriction of f_j to the set $T \setminus \mathbb{Q}$ is the constant function $c(t) \equiv \beta_j y$ on $T \setminus \mathbb{Q}$, whence $V_\varepsilon(f_j - f_k, T \setminus \mathbb{Q}) = 0$ for all $\varepsilon > 0$. Hence Theorem 4.25 is applicable to $\{f_j\}$. □

More examples, which can be adapted to the situation under consideration, can be found in [33, Section 4].

The following theorem is a counterpart of Theorem 4.14.

Theorem 4.28 *Let $T \subset \mathbb{R}$, $(M, \|\cdot\|)$ be a normed linear space, and $\{f_j\} \subset M^T$ be a pointwise relatively compact (or a.e. relatively compact) on T sequence of functions satisfying the condition: for every $p \in \mathbb{N}$ there is a measurable set $E_p \subset T$ with Lebesgue measure $\mathscr{L}(E_p) \leq 1/p$ such that*

$$\limsup_{j,k \to \infty} V_\varepsilon(f_j - f_k, T \setminus E_p) < \infty \quad \text{for all} \quad \varepsilon > 0.$$

Then $\{f_j\}$ contains a subsequence which converges almost everywhere on T.

Finally, we present an extension of Theorem 4.18 in the spirit of Theorems 4.15 and 4.16.

Theorem 4.29 *Let $T \subset \mathbb{R}$ and $(M, \|\cdot\|)$ be a reflexive Banach space with separable dual $(M^*, \|\cdot\|)$. Suppose the sequence of functions $\{f_j\} \subset M^T$ is such that*

(i) $\sup_{j \in \mathbb{N}} \|f_j(t_0)\| \leq C_0$ *for some $t_0 \in T$ and $C_0 \geq 0$;*
(ii) $\limsup_{j,k \to \infty} V_\varepsilon(\langle (f_j - f_k)(\cdot), x^* \rangle, T) < \infty$ *for all $\varepsilon > 0$ and $x^* \in M^*$.*

Then, there is a subsequence of $\{f_j\}$, again denoted by $\{f_j\}$, and a function $f \in M^T$ such that $f_j(t) \xrightarrow{w} f(t)$ in M as $j \to \infty$ for all $t \in T$.

References

1. Appell, J., Banaś, J., Merentes, N.: Bounded Variation and Around. Ser. Nonlinear Anal. Appl., vol. 17. De Gruyter, Berlin (2014)
2. Aumann, G.: Reelle Funktionen. Springer, Berlin (1954)
3. Aye, K.K., Lee, P.Y.: The dual of the space of functions of bounded variation. Math. Bohem. **131**(1), 1–9 (2006)
4. Banaś, J., Kot, M.: On regulated functions. J. Math. Appl. **40**, 21–36 (2017)
5. Barbu, V., Precupanu, Th.: Convexity and Optimization in Banach Spaces, 2nd edn. Reidel, Dordrecht (1986)
6. Belov, S.A., Chistyakov, V.V.: A selection principle for mappings of bounded variation. J. Math. Anal. Appl. **249**(2), 351–366 (2000); and comments on this paper: J. Math. Anal. Appl. **278**(1), 250–251 (2003)
7. Belov, S.A., Chistyakov, V.V.: Regular selections of multifunctions of bounded variation. J. Math. Sci. (NY) **110**(2), 2452–2454 (2002)
8. Bridger, M.: Real Analysis: A Constructive Approach. Wiley, Hoboken (2012)
9. Chanturiya, Z.A.: The modulus of variation of a function and its application in the theory of Fourier series. Soviet Math. Dokl. **15**(1), 67–71 (1974)
10. Chanturiya, Z.A.: Absolute convergence of Fourier series. Math. Notes **18**, 695–700 (1975)
11. Chistyakov, V.V.: The Variation (Lecture Notes). University of Nizhny Novgorod, Nizhny Novgorod (1992) (in Russian)
12. Chistyakov, V.V.: On mappings of bounded variation. J. Dyn. Control Syst. **3**(2), 261–289 (1997)
13. Chistyakov, V.V.: On the theory of multivalued mappings of bounded variation of one real variable. Sb. Math. **189**(5/6), 797–819 (1998)
14. Chistyakov, V.V.: On mappings of bounded variation with values in a metric space. Russ. Math. Surv. **54**(3), 630–631 (1999)
15. Chistyakov, V.V.: Mappings of bounded variation with values in a metric space: generalizations. J. Math. Sci. (NY) **100**(6), 2700–2715 (2000)
16. Chistyakov, V.V.: Generalized variation of mappings with applications to composition operators and multifunctions. Positivity **5**(4), 323–358 (2001)
17. Chistyakov, V.V.: On multi-valued mappings of finite generalized variation. Math. Notes **71**(3/4), 556–575 (2002)
18. Chistyakov, V.V.: Selections of bounded variation. J. Appl. Anal. **10**(1), 1–82 (2004)
19. Chistyakov, V.V.: Lipschitzian Nemytskii operators in the cones of mappings of bounded Wiener φ-variation. Folia Math. **11**(1), 15–39 (2004)

20. Chistyakov, V.V.: The optimal form of selection principles for functions of a real variable. J. Math. Anal. Appl. **310**(2), 609–625 (2005)
21. Chistyakov, V.V.: A selection principle for functions of a real variable. Atti Sem. Mat. Fis. Univ. Modena e Reggio Emilia **53**(1), 25–43 (2005)
22. Chistyakov, V.V.: A selection principle for functions with values in a uniform space. Dokl. Math. **74**(1), 559–561 (2006)
23. Chistyakov, V.V.: A pointwise selection principle for functions of a real variable with values in a uniform space. Sib. Adv. Math. **16**(3), 15–41 (2006)
24. Chistyakov, V.V.: Metric modulars and their application. Dokl. Math. **73**(1), 32–35 (2006)
25. Chistyakov, V.V.: Modular metric spaces, I: Basic concepts. Nonlinear Anal. **72**(1), 1–14 (2010)
26. Chistyakov, V.V.: Modular metric spaces, II: Application to superposition operators. Nonlinear Anal. **72**(1), 15–30 (2010)
27. Chistyakov, V.V.: Metric Modular Spaces: Theory and Applications. Springer Briefs in Mathematics. Springer, Cham (2015)
28. Chistyakov, V.V.: Asymmetric variations of multifunctions with application to functional inclusions. J. Math. Anal. Appl. **478**(2), 421–444 (2019)
29. Chistyakov, V.V., Chistyakova, S.A.: The joint modulus of variation of metric space valued functions and pointwise selection principles. Stud. Math. **238**(1), 37–57 (2017)
30. Chistyakov, V.V., Chistyakova, S.A.: Pointwise selection theorems for metric space valued bivariate functions. J. Math. Anal. Appl. **452**(2), 970–989 (2017)
31. Chistyakov, V.V., Galkin, O.E.: On maps of bounded p-variation with $p > 1$. Positivity **2**(1), 19–45 (1998)
32. Chistyakov, V.V., Galkin, O.E.: Mappings of bounded Φ-variation with arbitrary function Φ. J. Dyn. Control Syst. **4**(2), 217–247 (1998)
33. Chistyakov, V.V., Maniscalco C.: A pointwise selection principle for metric semigroup valued functions. J. Math. Anal. Appl. **341**(1), 613–625 (2008)
34. Chistyakov, V.V., Maniscalco C., Tretyachenko, Yu.V.: Variants of a selection principle for sequences of regulated and non-regulated functions. In: De Carli, L., Kazarian, K., Milman, M. (eds.) Topics in Classical Analysis and Applications in Honor of Daniel Waterman, pp. 45–72. World Scientific Publishing, Hackensack (2008)
35. Chistyakov, V.V., Nowak A.: Regular Carathéodory-type selectors under no convexity assumptions. J. Funct. Anal. **225**(2), 247–262 (2005)
36. Chistyakov, V.V., Repovš D.: Selections of bounded variation under the excess restrictions. J. Math. Anal. Appl. **331**(2), 873–885 (2007)
37. Chistyakov, V.V., Rychlewicz, A.: On the extension and generation of set-valued mappings of bounded variation. Stud. Math. **153**(3), 235–247 (2002)
38. Chistyakov, V.V., Tretyachenko, Yu.V.: A pointwise selection principle for maps of several variables via the total joint variation. J. Math. Anal. Appl. **402**(2), 648–659 (2013)
39. Cichoń, K., Cichoń, M., Satco, B.: On regulated functions. Fasc. Math. **60**(1), 37–57 (2018)
40. Dieudonné, J.: Foundations of Modern Analysis. Academic Press, New York (1960)
41. Di Piazza L., Maniscalco, C.: Selection theorems, based on generalized variation and oscillation. Rend. Circ. Mat. Palermo, Ser. II, **35**(3), 386–396 (1986)
42. Dudley, R.M., Norvaiša, R.: Differentiability of Six Operators on Nonsmooth Functions and p-Variation. Lecture Notes in Math., vol. 1703. Springer, Berlin (1999)
43. Fraňková, D.: Regulated functions. Math. Bohem. **116**(1), 20–59 (1991)
44. Fraňková, D.: Regulated functions with values in Banach space. Math. Bohem. **144**(4), 1–20 (2019)
45. Fuchino, S., Sz. Plewik, Sz.: On a theorem of E. Helly. Proc. Am. Math. Soc. **127**(2), 491–497 (1999)
46. Gniłka, S.: On the generalized Helly's theorem. Funct. Approx. Comment. Math. **4**, 109–112 (1976)
47. Goffman, C., Moran, G., Waterman, D.: The structure of regulated functions. Proc. Am. Math. Soc. **57**(1), 61–65 (1976)

48. Goffman, C., Nishiura, T., Waterman, D.: Homeomorphisms in Analysis. Amer. Math. Soc., Providence (1997)
49. Helly, E.: Über lineare Funktionaloperationen. Sitzungsber. Naturwiss. Kl. Kaiserlichen Akad. Wiss. Wien **121**, 265–297(1912) (in German)
50. Hermes, H.: On continuous and measurable selections and the existence of solutions of generalized differential equations. Proc. Am. Math. Soc. **29**(3), 535–542 (1971)
51. Hildebrandt, T.H.: Introduction to the Theory of Integration. Academic Press, New York, London (1963)
52. Honig, C.S.: Volterra Stieltjes-integral Equations: Functional Analytic Methods, Linear Constraints (Mathematics Studies). North-Holland, Amsterdam (1975)
53. Łojasiewicz, S.: An Introduction to the Theory of Real Functions, 3rd edn. Wiley, Chichester (1988)
54. Maniscalco, C.: A comparison of three recent selection theorems. Math. Bohem. **132**(2), 177–183 (2007)
55. Medvedev, Yu.T.: A generalization of a theorem of F. Riesz. Usp. Mat. Nauk **8**(6), 115–118 (1953) (in Russian)
56. Megrelishvili, M.: A note on tameness of families having bounded variation. Topol. Appl. **217**(1), 20–30 (2017)
57. Menger, K.: Untersuchungen über allgemeine Metrik. Math. Ann. **100**, 75–163 (1928)
58. Musielak, J.: Orlicz Spaces and Modular Spaces. Lecture Notes in Math., vol. 1034, Springer, Berlin (1983)
59. Musielak, J., Orlicz, W.: On generalized variations (I). Stud. Math. **18**, 11–41 (1959)
60. Musielak, J., Orlicz, W.: On modular spaces. Stud. Math. **18**, 49–65 (1959)
61. Perlman, S.: Functions of generalized variation. Fund. Math. **105**, 199–211 (1980)
62. Natanson, I.P.: Theory of Functions of a Real Variable, 3rd edn. Nauka, Moscow (1974) (in Russian); English translation: Ungar, New York (1965)
63. Ramsey, F.: On a problem of formal logic. Proc. Lond. Math. Soc. (2) **30**, 264–286 (1930)
64. Rao, M.M.: Measure Theory and Integration. Wiley, New York (1987)
65. Riesz, F.: Untersuchungen über Systeme integrierbarer Funktionen. Ann. Math. **69**, 449–497 (1910)
66. Rudin, W.: Principles of Mathematical Analysis, 2nd edn. McGraw-Hill Book, New York (1964)
67. Saks, S.: Theory of the Integral. Second Revised Edition. Stechert, New York (1937)
68. Schrader, K.: A generalization of the Helly selection theorem. Bull. Am. Math. Soc. **78**(3), 415–419 (1972)
69. Schramm, M.: Functions of Φ-bounded variation and Riemann-Stieltjes integration. Trans. Am. Math. Soc. **287**, 49–63 (1985)
70. Schwabik, Š.: Generalized Ordinary Differential Equations. World Scientific, River Edge (1992)
71. Schwabik, Š.: Linear operators in the space of regulated functions. Math. Bohem. **117**(1), 79–92 (1992)
72. Schwartz, L.: Analyse Mathématique, vol. 1. Hermann, Paris (1967) (in French)
73. Talvila, E.: The regulated primitive integral. Ill. J. Math. **53**(4), 1187–1219 (2009)
74. Tret'yachenko, Yu.V.: A generalization of the Helly theorem for functions with values in a uniform space. Russ. Math. (Iz. VUZ) **54**(5), 35–46 (2010)
75. Tret'yachenko, Yu.V., Chistyakov, V.V.: The selection principle for pointwise bounded sequences of functions. Math. Notes **84**(4), 396–406 (2008)
76. Tvrdý, M.: Regulated functions and the Perron-Stieltjes integral. Časopis Pěst. Mat. **114**, 187–209 (1989)
77. Tvrdý, M.: Generalized differential equations in the space of regulated functions (boundary value problems and controllability). Math. Bohem. **116**(3), 225–244 (1991)
78. Waterman, D.: On convergence of Fourier series of functions of generalized bounded variation. Stud. Math. **44**, 107–117 (1972)
79. Waterman, D.: On Λ-bounded variation. Stud. Math. **57**, 33–45 (1976)

Index

© The Author(s), under exclusive license to Springer Nature Switzerland AG 2021
V. V. Chistyakov, *From Approximate Variation to Pointwise Selection Principles*,
SpringerBriefs in Optimization, https://doi.org/10.1007/978-3-030-87399-8

Printed in the United States
by Baker & Taylor Publisher Services